ENGINEERING—AN
ENDLESS FRONTIER

ENGINEERING—AN
ENDLESS FRONTIER

Sunny Y.
Auyang

Harvard University Press

Cambridge, Massachusetts & London, England

2004

Library of Congress Cataloging-in-Publication Data

Auyang, Sunny Y.
Engineering : an endless frontier / Sunny Y. Auyang.
   p. cm.
Includes bibliographical references and index.
ISBN 0-674-01332-8 (alk. paper)
1. Engineering.   I. Title.

TA157.A96 2004
620—dc22   2003057123

CONTENTS

PREFACE

Talking on the phone while cruising down the freeway at sixty-five miles per hour, we take for granted the technologies that make possible many of our daily activities. Yet perhaps in some rare moment, parked at a desolate place and looking at the tiny device by which you've just talked to a loved one hundreds of miles away, questions have flashed through your mind: What does it mean to me? How does it work? How do they make it? *They* are mostly engineers. This book explains how engineers create technology. A big picture of modern engineering reveals that it has both physical and human dimensions. Besides designing useful products, engineers are also adept in science and management, both crucial to technological progress.

I am a physicist who has turned from doing physics to thinking about science and technology. My graduate school and physics career, mostly at MIT, brought me many engineer friends. In writing this book, I sat in on many classes and had the pleasure of learning about many fascinating topics. I am grateful to the professors and other helpful engineers. In particular, I thank Nicholas Ashford, Vincent Chan, Millie Dresselhaus, Albert Kündig, Paul Lagace, Richard Lester, and Horace Yuen for stimulating conversations. The manuscript has also benefited from many suggestions by Michael Fisher of Harvard University Press.

To make this book enjoyable to a wide range of readers, I have used the concise style standard in scientific references and have avoided topics of only scholarly interest. Readers who wish to dig deeper can find an extended bibliography, additional information, and certain scholastic qualifications on the website www.creatingtechnology.org.

1

INTRODUCTION

"The engineer typifies the twentieth century. Without his genius and the vast contributions he has made in design, engineering, and production on the material side of our existence, our contemporary life could never have reached its present standard." Thus wrote Alfred Sloan, who, with a college degree in engineering, climbed the white-collar career path to the presidency of General Motors. Engineering not only elevates society's material culture but also makes for satisfying personal lives, as observed by Herbert Hoover, a mining engineer whose life career culminated as president of the United States: "It is a great profession. There is the fascination of watching a figment of the imagination emerge through the aid of science to a plan on paper. Then it moves to realization in stone or metal or energy. Then it brings jobs and homes to men. Then it elevates the standards of living and adds to the comforts of life. That is the engineer's high privilege."[1]

Engineering is the art and science of production that, alongside reproduction, is the most fundamental of human activities. As practical art, it flourished with the rise of civilization. Even in ruins, monuments around the world display the ingenuity of ancient master builders. Modern engineering, which emerged together with modern science in the seventeenth-century scientific revolution, amplifies traditional ingenuity by the power of scientific reasoning and knowledge. Directly involved in material production, it acts at the vortex, merging research and development, on the one hand, and industry and business, on the other. Science, engineering, and business jointly constitute the main engines of technology, whose products have been woven into the fabrics of our daily life. This book focuses on engineering, not neglecting its overlaps and bonds with the other two. What are the characteristics of engineering thinking and practice that make them so effective in advancing technology? How are they related to those of natural science? What factors do engineers reckon with in designing a system? How do their decisions shape technological evolution?

As a creative and scientific activity that transforms nature to serve the needs and wants of large numbers of people, engineering has both physical and human dimensions. To modify nature effectively as desired requires mastery of natural laws and phenomena, thus engineering shares the contents and standards of natural science. To ascertain what modifications are desirable requires an understanding of human and socioeconomic factors, thus engineering goes beyond natural science in its missions of utility and service. Engineers address things and people, bringing nature and humanity together. They develop physical technology to grasp general principles, extract materials, create implements, design products, and invent processes for efficient manufacturing and operation. They also develop organizational technology to analyze goals, assess costs and benefits, trade off factors under constraints, negotiate consensus among involved parties, plan projects, and coordinate human efforts in production. The physical and organizational dimensions infiltrate and reinforce each other in the advancement of engineering, technology, and social well-being.

Engineering has progressed remarkably during the past century. Its scientific sophistication soars with the complexity of technologies it creates. Its management and social horizon expand with the scope of production and technological risks. Exercising the physical and human arms of their profession, engineers engage in science, design, and leadership. Through these activities they improve knowledge, implements, and organizations, which are the major repositories of technology, society's scientific productive capacity.

Engineering is scientific. Its first aspect, science, is closely related but not identical to natural science. The relation between the two was aptly described by electrical engineer Vannevar Bush, whose 1945 report to the U.S. president, entitled *Science—The Endless Frontier,* set the vision of the National Science Foundation (NSF). He wrote elsewhere: "In all associations between engineers and scientists, engineering is more a partner than a child of science . . . While everyone knows that engineering is concerned with the conversion of science into technology, everyone does not know that engineering also does just the opposite and translates technology into new science and mathematics."[2] Scientists know. Physicist James Clerk Maxwell acknowledged in his treatise that revolutionized electromagnetic theory: "The important applications of electromagnetism to telegraphy have also reacted upon pure science by giving a commercial value to accurate electrical measurements, and by affording to electricians the use of apparatus on a scale which greatly transcends that of any ordinary laboratory. The consequences of this demand for electrical knowledge, and of these experimental opportunities for acquiring it, have been already very great, both in stimulating the energies of advanced electricians, and in diffusing among practical men a degree of accurate knowledge which is likely to conduce to the general scientific progress of the whole engineering profession."[3] That was 1873. Since then the partnership of science and engineering has strengthened so tremendously that science and engineering are often treated as a unit by the NSF.

Natural scientists discover what was not known. Engineers create what did not exist. Both boldly go where no one has gone before, each original and creative in its own way. Through intensive research efforts in the past fifty years, engineers have developed engineering sciences, bodies of coherent knowledge that are comparable to the natural sciences in their length of vision, breadth of scope, depth of analysis, level of creativity, rigor of research, and criteria of acceptance. In the process they are neither imitating natural scientists nor ignoring them. Engineers mainly develop and refine the independent reasoning inherent in productive activities, which both Aristotle and Galileo analyzed and treated on equal footing as reasoning in pure study. Independence is not insularity. Engineers and natural scientists may have different motivations and concerns, but they share the human mind and the physical world. They have common knowledge, methods, and ways of thinking, which include mathematics and instrumentation, theory formation and controlled experimentation, research and development. In many workplaces they collaborate so closely it is difficult to tell them apart. Their domains increasingly overlap as swelling knowledge bursts the boundaries of academic departments and society increasingly demands useful results from research. Genetic engineering occupies much of cutting-edge biology, and nanotechnology looms as a direction for future chemistry. Convergence accelerates and interdisciplinary research centers mushroom as the twenty-first century dawns.

While engineering science expands knowledge, design creates specific implements and systems, ranging from the goods and services that people use every day to society's infrastructures for energy, communication, transportation, public health, and national defense. Design is not unique to engineering, but its paramount importance is. Design of good experiments is crucial to natural science, but there it is at the service of the central goal of acquiring knowledge. The emphasis is reversed in engineering, where design is the central goal that calls on the service of scientific knowledge. Engineering design is creative and imaginative, although invariably constrained by many conditions that the intended product is required to meet. It includes processes from the conception of a system through development to the plan for producing it. In the process engineers must attend to the performance of the system as a whole as well as its components down to the smallest detail, not neglecting the erratic environment in which it is intended to operate. To ensure that the product works reliably and safely, they try to anticipate problems and head them off. No scientific principle can cover all the vicissitudes of reality. Thus design engineers must complement their mathematical prowess and explicit knowledge with practical insight and know-how. Insights and experiences originated in design practice, when scrutinized, developed, and systematized, become new scientific engineering knowledge. In his seminal study in the epistemology of design, aeronautic engineer Walter Vincenti remarks that a great number of engineers "engage in design, and it is there that requirements for much engineering knowledge originate in an immediately technical sense."[4]

Design is akin to invention. In creating technology, however, engineers do not merely invent, they innovate. Innovation, which includes the whole process from discerning the need for an invention to bringing the new idea through design and production to large-scale use and economic return, has a broad perspective that requires both technical and social acumen. Consider designing and building a subway system, for instance. Even when funding is available, the communities to be served—or inconvenienced—are heterogeneous populations, not to mention the politicians, environmentalists, and myriad other interest groups involved. Engineers must listen to and negotiate with them all, and the outcomes will influence the design of the system, such as the locations of stations. Then the construction process brings in more parties, from workers and contractors to drivers cursing traffic detours. The importance of organizational technology is unmistakable.

Engineered systems are real and have real consequences. Engineers have heavy social responsibilities to manage intricate technologies that, while elevating living conditions, also carry risks for accidents, abuses, and harmful side effects. To meet the challenge, they expand their horizon to cover societal needs, government regulations, and impacts on the environment. It falls on them to penetrate media hype and ideological rhetoric to explain realistic factors clearly to the public, thus to help people to evaluate relative importance, weigh risks, make trade-offs, and decide rationally on public policies concerning technology. For these tasks engineering schools are intensifying education in social and communicative skills. Many engineers agree with their colleague Norman Augustine: "Engineers must become as adept in dealing with societal and political forces as they are with gravitational and electromagnetic forces."[5]

The dual physical and human dimensions and triple aspects of science, design, and leadership provide the framework for our exploration of engineering. This multidimensional framework does not imply that each engineer is expert in all three aspects. Engineering is a large and diverse field. There are more than two million working engineers and more than a hundred professional societies in the United States tending to mechanical, information, and other kinds of technology. Overwhelming technological complexity makes the division of intellectual labor inevitable. Thus engineer-scientists focus on research, engineer-entrepreneurs and engineer-managers on organization. The majority of engineers focus on design and development, which many regard as the core of engineering. The focus, however, is not tight. An engineer working in one aspect of engineering is aware of the other two, although not with deep expertise. Engineer-scientists are not confined to ivory towers, design engineers are not mere technicians, and engineer-managers are not simply bean counters. They are all engineers concerned about creating and advancing technology. The broad view of engineering is aptly described by civil engineer Frederick Clarke:

> It involves the delicate and difficult task of translating scientific abstraction into the practical language of earthly living; and this is perhaps the most completely demanding task in the world. For it requires an understanding of *both* spheres—the pure ether in which science lives, and also the goals and drives and aspirations of human society in all its complexity. The engineer must be at once philosopher, humanist and hard-headed, hard-handed worker. He must be a philosopher enough to know what to *believe*, humanist enough to know what to *desire*, and a workman enough to know what to *do*.[6]

This book presents a big picture of engineering. Big pictures of complex topics are necessarily sketchy and incomplete. No single work can cover, even in outline, such a rich field as engineering with its numerous branches that attend to all corners of the technological world: aeronautic and astronautic, agricultural, automotive, biological and biomedical, chemical, civil and structural, computer and software, electrical and electronic, environmental, industrial, marine, material, mechanical, mining and petroleum, nuclear, systems, and many more. To distill their general characteristics without leaving out all their details, I rely heavily on the integration of concepts and cases.

The book comprises two parts, each with three chapters. Chapters 2, 3, and 4 examine engineering's history and socioeconomic contexts. Chapters 5, 6, and 7 analyze its content and contributions to technology. The two halves of the book separately employ a case-exemplifying-concept approach and a concept-illustrated-by-case approach. Narrative dominates the historical account, but historical events are often presented as cases that exemplify engineering concepts, which are dispersed among the stories. For instance, the emergence of chemical engineering is described as exemplifying how engineering bridges science and industry. The content part explains the overarching concepts of each topic in a synoptic view, then immediately illustrates them with cases chosen from various branches of engineering. For instance, the structure of systems engineering from requirement elicitation to detail design is analyzed and illustrated by two aircraft-development projects. In this way, we see both an overall picture of engineering and the specialties of each branch. The particular examples illuminate the general concepts at issue and simultaneously gain a broader significance.

The history chapters include intellectual and social strands. The intellectual history in Chapters 2 and 3 starts with the historical evolution of the concepts of technology, engineering, and science. For the history of engineering proper, I concentrate on four major branches: civil, mechanical, chemical, and electrical and computer. In each branch I try to delineate general themes that are shared by other branches, for instance mathematical analysis and rational empiricism, building scientific foundations and building new industries. I hope they convey a sense of the development of engineering at large.

Chapter 4 describes the rising social status of engineers and their organizations. Beginning with self-educated pioneers, it traces the emergence of professional societies, undergraduate and graduate education, research universities, and industrial laboratories. Against the background of the U.S. economy, it explains the contributions of engineering and their influences on industrial organizations.

The three chapters in the second half of the book, which delves into the technical and professional content of engineering, separately focus on engineering's aspects of design, science, and leadership. Chapter 5 begins with an analysis of several general types of thinking shared by engineers and natural scientists. Then it looks at design on two scales. On a small canvas it presents several cases showing the heuristics, judgments, and justifications in the thinking of individual inventors. On a large canvas it presents the design principles in systems engineering with its extended view of technological systems. Individual thinking is intimate but often unstated and intuitive. Systems engineering comprises sound common sense articulated, systematized, synthesized, and brought to bear on complex situations. Originally developed to facilitate large projects, its ideas are equally useful for individual design engineers.

Chapter 6 presents engineering sciences as coherent bodies of knowledge and considers such epistemological questions as why mathematics is so effective here. Engineering sciences fall roughly into two types: *systems,* such as information theory, and *physical,* such as fluid dynamics. Systems sciences are indigenous to engineering. Engineering sciences in the physical group contain novel concepts absent in their natural-science counterparts. One strength of engineering sciences is their integrative power, by which they synthesize principles from various academic disciplines and bring them to bear on a wide class of practical problems. They are crucial for the future development of nanotechnology and biological engineering.

Chapter 7 explores how engineers engage in business and reach out to society, what ethical problems they face, and why technical factors are often dwarfed by other considerations. Here the human dimension of their profession comes into focus. Engineers play important roles in the management of private enterprises and lesser but not negligible roles in shaping public policy. As custodians of risky technologies, they are under increasing social and professional obligations to design for reliability, safety, environmental benignity, and sustainability. These tasks become more challenging as living standards rise for greater portions of the world's expanding population and natural resources are finite. So long as sustainable development is a worthy goal, the frontier of engineering is endless.

2

TECHNOLOGY TAKES OFF

2.1 From Practical Art to Technology

"We have three mirrors. The mirror of bronze brings knowledge about attire; the mirror of other people, about strengths and weaknesses; the mirror of history, about rise and fall of powers," wrote a seventh-century emperor during whose reign China reached a peak in culture and prosperity. The mirror metaphor underscores the importance of reflection. The Delphic injunction *know thyself* is limited neither to the Greeks nor to antiquity. It is "leadership's first commandment," according to a recent article on business administration.[1] A good way to know yourself is to reflect on what you have gone through. Engineers often perform a retrospective analysis at the end of a project to draw out lessons learned, and their colleagues consult these on embarking on new endeavors. In a similar spirit, history is studied here not as an academic pastime but as a means to draw on past experiences to understand the present and engineer the future.

Today, engineers are peers of natural scientists in research and leaders in development. Engineering, not including computer science, receives about a tenth of U.S. federal funding in basic research, a quarter in applied research, and the lion's share in development. Its position in industry-funded research is more favorable because of its practicality.

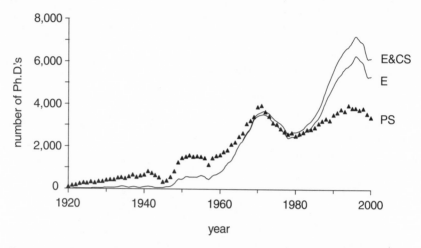

Figure 2.1 Doctorate degrees awarded each year by U.S. universities in engineering (E), engineering and computer science (E&CS), and the physical sciences (PS). *Source:* Census Bureau, *Historical Statistics of the United States* (1975), p. 387; National Science Foundation, *Science and Engineering Degrees* (2002), tables. 19, 46.

American universities award more doctorates in engineering than in mathematics and physical sciences combined (Fig. 2.1). In its climb to this status, what capabilities has engineering developed?

Technology as the Development of Productive Activity
The intellectual ascent of engineers would not have surprised Socrates, who consulted their forerunners, practical artists, and acknowledged: "They understood things which I did not, and to that extent they were wiser than I was." Just as we now talk about science and technology in one breath, the ancient Greeks mentioned *epistēmē* (narrowly science and broadly knowledge in general) and *téchnē* (art) together, sometimes using the two terms interchangeably.[2]

"I will not call *téchnē* anything without *logos*," declared Plato. *Logos*, broadly meaning language, reasoning, and rational account, was an important concept in Greek philosophy. The Aristotelian dictum that through Latin is known to us as "man is a rational animal" was literally "an animal with *logos*." Plato explained that by virtue of its possession of *logos*, a *téchnē* such as medicine was distinct from *empeiria* (experience, knack) such as cooking.[3]

The distinction between art and experience was elucidated by Aristotle. He looked at the human mind as having two parts, an irrational part without *logos* and a rational part with *logos*. Further analyzed, the rational mind reveals at least three important capacities. Science *(epistēmē)* addresses the topic of unchanging being and underlies the activity of contemplation *(theōria)*. Art *(téchnē)* addresses bringing into being and underlies production *(poiēsis)*. Prudence *(phronēsis)* addresses conduct and underlies action *(praxis)*. All three mental faculties yield knowledge as opposed to mere opinions. The articulated principles and knowledge in theoretical and prudential activities have been generally called science and ethics. No general name existed for the principles and knowledge in productive activities until the nineteenth century and the emergence of technology.

"*Téchnē* is a state of capacity to produce with a true *logos*," Aristotle stated, and he proceeded to explain how artists acquire the capacity to produce with a true course of reasoning: "Art arises, when from many notions gained by experience one universal judgment about similar objects is produced . . . We think that knowledge and understanding belong to art rather than to experience, and we suppose artists to be wiser than men of experience and this because the former know the cause, but the latter do not. For men of experience know that the thing is so, but do not know why, while artists know the 'why' and the cause of thing that is done . . . And in general it is a sign of the man who knows, that he can teach, and therefore we think art more truly knowledge than experience is, for artists can teach, and men of mere experience cannot." Art included many arts: mathematics, rhetoric, medicine, architecture, and more. The artist in architecture, the master builder, distinguished himself from laborers by his knowledge of the causes of construction. Aristotle analyzed *cause* into four aspects: material, formal, efficient, and final causes. Scholars have pointed out that his "cause" was actually "because" and his four causes answered four *why* questions: because it is made of a certain material; because its material is organized in a certain form; because it is driven by a certain force; because it is for the sake of a certain purpose.[4] Aristotle expected artists to be able to give correct material, structural, dynamic, and functional explanations of their works. Whether or not they lived up to the expectation, our engineers do.

As the ability of creative production, *téchnē* included not only practical arts but also fine arts and literature, where activities with reasoned considerations bring forth paintings and poems. Both fine and practical arts can be practiced by the same person, a phenomenon most pronounced in the Renaissance. Historian Anthony Grafton observed: "They [Renaissance engineers] made no fine distinctions between what we would now call the fine arts, like painting and sculpture, and the applied ones, like bridge building. On the contrary, the same masters often carried on projects in all these areas."[5] He described how engineers such as Leonardo da Vinci excelled in both fine and practical arts and did not acknowledge any inferiority of one to the other. Socially, practical and fine arts had acquired plebeian and patrician images, respectively, because one belonged more to common people and the other to the upper classes. However, philosophers in ancient Greece, who were not ensconced in ivory towers but talked to people in the marketplaces, realized that intelligence resides not in social ranks. Although productive and theoretical thinking take on different topics that may be subjected to social discrimination, their intellectual qualities are equal. With slight modification, Aristotle's analysis of *téchnē* quoted earlier applies also to scientific generalization and theory formation. It points to the conception of technology now familiar to engineers and natural scientists.

Téchnē contains its own *logos*, which is systematically advanced in technology. To produce the means of living, find meanings of life in work, coordinate efforts, fashion tools, use resources, and modify nature to make it more hospitable constitute common human capabilities necessary for survival. The development of these capabilities into modern technology is like the growth of an acorn into an oak tree; the growth principle is mainly its own, although it also absorbs information from and exerts influences on natural science and industry. Far from being merely a collection of mindless physical and social systems, technology is inherently intellectual, being the articulation, generalization, refinement, and systematization of reasoning and knowledge involved in productive and creative activities that are at the core of human existence.

Possession of articulated scientific principles is a hallmark of technology, which differs from traditional practical art as modern physics differs from Greek astronomy. It is no surprise that the concept of technology spread concomitantly with the rise of modern engineers and applied scientists who increasingly developed and articulated their own reasoning. In 1829, botanist Jacob Bigelow delivered a series of lectures on technology, defining it as "the principle, processes, and nomenclatures of the more conspicuous arts, particularly those which involve application of science, and which may be considered useful, promoting the benefit of society, together with the emolument of those who pursue them."[6] He sat on the board of trustees of Massachusetts Institute of Technology (MIT), the foundation of which in 1861 helped to popularize the notion of technology. Unlike apprentices in workshops who are trained in a skill that is poorly articulated, students in universities gain their experiences in scientific laboratories that simultaneously impart generalizable knowledge and nurture independent reasoning. Realizing the advantage of systematic articulation for mass education, institutes of higher education replaced workshops as the major grounds of learning for practical artists—or engineers, as they were increasingly called.

Predecessors of Modern Engineers

Modern engineers have a proud heritage. Their forerunners contributed greatly to the prosperity of their contemporaries. By constructing roads, bridges, canals, hydraulic devices, and water and sanitary systems, they secured production of material means for living well. By erecting monumental buildings, they gave physical manifestation and inspiration to their culture's collective imagination and spiritual aspiration. Many of their works have stood through the ravages of time to inspire awe: the pyramids of Egypt and Mesoamerica, the great wall and grand canal of China, Roman aqueducts and the Colosseum, the Taj Mahal of India.

Forerunners of engineers came from various backgrounds. A few, such as Archimedes, were also physicists and mathematicians. Others were scholarly officials who, being put in charge of building projects, studied their subject and made contributions. An example was Frontinus, who supervised the construction of the Roman aqueducts and wrote one of the two books that survive to tell us about the details of Roman engineering. Many others were skillful artisans who rose to the top of their craft.[7]

Ancient master builders were broadly knowledgeable. The Greek *architekton* or Roman *architectus* (master builder) who designed and directed the construction of harbors, water supplies, and public buildings, was at once what we now call the architect, adept in design and function; the civil engineer, proficient in surveying and structural integrity; the mechanical engineer, familiar with hoisting and lifting machinery; and the manager, skillful in organizing workers. So was the master mason of medieval Europe. Vitruvius's *De Architectura*, written around 20 B.C.E., described the ideal master builder as being in command of written language, drawing, geometry, arithmetic, and optics, and not ignorant of astronomy, history, philosophy, music, medicine, and law. Astronomy provided the backdrop for surveying, history for ornamentation in design, and medicine for planning water supplies and city layouts. Familiarity with local laws enabled proper and smooth construction of civic structures. Knowing music enabled one to tune, on strength of sound, the tension in ropes of hoists and catapults.[8]

Vitruvius asserted in the opening of his treatise that master builders were ingenious, or in possession of *ingenium*. People's admiration for their resourcefulness was expressed by the name they gave them. From the eleventh century, builders of ingenious devices and fortifications—for example, Ailnolth (fl. 1157–1190), who worked on the Tower of London—were called *ingeniator* in Latin. Through the French *ingénieur*, it later became the English *engineer*. The designation was significant in two ways. First, it differed from such titles as "baker" or "silversmith," which distinguished a worker by the material he worked with. The *ingeniator* was distinguished by his *ingenium*, which meant both ingenuity and the product of ingenuity. These meanings were preserved in the English term *engine*, which meant genius and ingenuity before being taken over by steam engine and its like—a case of creators overshadowed by their creations. Second, whereas names like "scientist" and "physicist" were coined by academicians for those previously called natural philosophers, "engineer" originated in everyday usage, thus conveying a sense of spontaneous appreciation.[9] You may have felt that same appreciation yourself if you have ever been awe-stricken by a Gothic cathedral in Northern Europe and wondered how it was built.

The name "ingeniator" acquired wide currency during the Renaissance, with the rise of a new breed of engineers who not only excelled in but were articulate about many arts. Grafton observed that the Italian Renaissance witnessed "a technological and social revolution—one that radically changed both the social structures that created buildings and works of art and the products that resulted. The protagonists of this revolution, as Jacob Burckhardt suggested long ago, were engineers."[10] Among Renaissance engineers was Leonardo, who was adroit in mathematics, studied machines of all kinds, supplied a long list of his engineering abilities on his resume, and for some time bore the title of Ingegnere Generale.[11]

In late seventeenth-century France, *ingénieur* became the title of formally educated technical officers. From there it was naturally adopted by the emerging scientific professionals. The first to emerge, who engaged mostly in construction, called themselves civil engineers. *Civil engineering* began as an inclusive term; "civil" meant civilian as distinct from military. Its meaning narrowed with the assertion of mining, mechanical, and other branches of engineering, whose concerns were not construction. Now many branches of engineering exist, each identified by the nature of their work.[12]

Engineering and Applied Science

From the charter of the first engineering society to today, engineers talk proudly of the application of science. Science generally is the state of knowing or possessing knowledge that is sufficiently general, clearly conceptualized, carefully reasoned, systematically organized, critically examined, and empirically tested. According to topics, it divides into natural sciences, such as physics and biology; engineering sciences, such as computation and information; and social and human sciences.

While craft tends to be absorbed by the specific task at hand, scientific thinking considers a general type of task with many varieties. Stereotypical craftsmen are content to know how things in their trade work but not why, to produce but not to conceptualize about the process of production, to use their own stock of knowledge on the problem at hand but not to acquire new knowledge actively. Craft inventions rely mostly on tinkering, varying slightly existing solutions but not venturing far, for those efforts would require too many resources, both intellectual and material. In contrast, engineers are kin to natural scientists, who are not content to be stuck to immediate tasks, although they never forget utility. Individually, an engineer may have to conform to a production schedule that prohibits comprehensive investigation. Collectively in a discipline, engineers are willing to spend additional resources for a broader perspective; to investigate why things work as they do; to question ready-made answers; to criticize their own convictions; to actively seek alternatives that are not apparent; to generalize, articulate, and disseminate their discoveries; to acquire new knowledge, some of which may not be directly relevant. These efforts expand vision but demand heavy overhead costs, just as going to college is expensive in lost income if not tuition. By stepping back from the immediate task and looking beyond it, the scientific vision consumes resources and may temporarily hurt craft performance. Thus the development of scientific engineering was painfully slow at the beginning, and scientific thinking was dismissed by contemporary critics and some historians as superfluous if not counterproductive. Nevertheless, when engineers discover the typical factors shared by diverse artifacts, they avail themselves of a wide range of possibilities for designing particular systems. The scientific mastery enables them to leapfrog to more complex systems, leaving craftsmen far behind.

Many engineers identify engineering with applied science, and schools of engineering and applied science thrive in many universities.[13] The distinction between pure and applied science is not a clear one. Perhaps the most often cited criterion is the motivation of researchers: curiosity and the urge to know, or utility and the urge to invent. This is unsatisfactory because human motivations are many and mixed. Elementary particle physics is perhaps the purest of sciences, yet physicists require huge accelerators and other instruments useful for their pursuit. Ludwig Prandtl and his mechanical engineering followers had flying machines in their vision while advancing knowledge in theoretical aerodynamics. Scientists and engineers, generally, differ less in the choice between knowledge and utility than the relative weights the two factors may have for a particular researcher. It is a fuzzy distinction with significant overlap, which increases in large projects that assemble many talents.

Disparate motivations do not dictate disparate approaches and results. Discoveries of wonder-motivated research are not necessarily useless; pure research on electromagnetism and countless other natural phenomena bear utilitarian fruits that drastically improve the ways we live and communicate. Nor is utility-motivated research necessarily kept on a short leash by immediate needs. Strategic engineering research, such as deep-sea drilling or harnessing nuclear fusion for emission-free energy, are like the research to fight cancer. They have utilitarian missions, but these missions are extremely complex and involve many poorly understood phenomena that must be studied scientifically. Many such utility-motivated studies are long-term enterprises requiring as much imagination and ingenuity as wonder-motivated research. Their results are not proprietary but published, judged not by client satisfaction but by peer review. These practices are not different from those in pure research.

The relation between pure and applied sciences is not hierarchical. One does not generally or necessarily depend on the other. Applied mathematics is neither the application of pure mathematics nor epistemologically dependent on it. It is a differently oriented mathematics that historically appeared before pure mathematics and is not intellectually inferior. Other applied disciplines are similar. Knowledge about nature can be gained through utilitarian activities or disinterested research. One can discover something new in pure research and then apply it for practical purposes, as in applications of quantum mechanics to electronic devices. One can also practice it first and then abstract the knowledge from intuition and systematize it, as with the development of thermodynamics after the spread of heat engines. Both phenomena have occurred countless times in the history of science and technology.[14] In short, the difference between pure and applied science is one of orientation and not intellectual quality. Prejudice and stereotypical distortion aside, engineering as applied science, a science oriented toward utility and application, is a legitimate and proud identity.[15]

Dynamics of Technological Change

Technology is a scientific capacity for creative production and an enabler of activites. Like all potentials, it may or may not be realized. A depressed economy with plants idled and engineers and scientists driv-

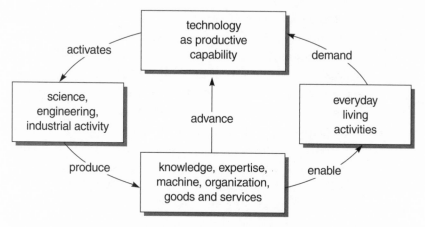

Figure 2.2   The dynamics of technology as a scientific productive capacity.

ing taxi cabs is wasting its technological capacity. Workers exercise a society's technological capacity in their technical activities. Activities in engineering, science, and industry produce results, which in turn generate demand pull and supply push for technological progress (Fig. 2.2).

Technological results come in many tangible and abstract forms. The bulk consist of a dazzling variety of goods and services that are consumed by people in their daily lives. Technology is conspicuous in devices that enable people to do things otherwise deemed "magical," such as talking to a friend thousands of miles away while lounging at the beach, or performing a medical diagnosis of internal organs with advanced imaging instruments. Even in foods and fabrics, where it is not conspicuous, technology quietly contributes to the production and distribution systems that make them widely available and affordable. Technological goods and services ameliorate material conditions of living, thereby opening new possibilities and enabling more people to have more choices in their ways of life. They do not come without costs, such as polluting the environment, dislocating jobs, and undermining customs. People can reject most if not all new technologies; some do and some advocate doing it. The vast majority of Americans, however, deem the costs small compared with what technology enables them to do. Even so, they do not passively accept everything.[16] Their active choices among different technological products generate differential demand pull that influences the direction of technological change.

The supply or creative side is the workplace of the other results of technological activity: tacit expertise and articulated knowledge, tools and machines, physical and social infrastructures. Compared with goods and services, they are small in quantity but not in consequence. Like profits reinvested for economic growth, they are plowed back in for technological progress. They expand the technological capacity by enlarging its four major repositories: knowledge, implements, organizations, and humans.

Knowledge in general can be explicit or implicit. Tacit knowledge is embedded in human skills and experiences, layouts and operations of physical implements and plants, and organizations of work and administration. I will refer to these in specific terms and use the word *knowledge* mainly for information expressly articulated. Explicit knowledge can be factual or procedural. Factual knowledge about states of affairs ranges from physical laws to the suitability of various materials for various constructions. Procedural knowledge about how to do things ranges from algorithms for computer sorting to catalytic cracking processes for oil refining. In principle, information written out is portable. In practice, technology transfer is difficult and uncertain; besides uncontrollable contingencies, it depends on conditions of the three other repositories.

The second repository of technology consists of artifacts and implements. Aside from myriad appliances used in daily life, it comprises scientific and medical instruments; machines and plants for agriculture and manufacture; buildings, roads, bridges, and ports; vehicles for sea, land, air, and space transport; pipelines and electricity grids; telephone and broadcast networks; water and waste treatment facilities. These are more than investments of material resources for production. Engineers distill certain experiences and skills and design them into hardware, making some implements to operate automatically, others to interface smoothly with human operators, and still others to amplify human abilities. Thus implements are physical embodiments of knowledge, know-how, and skills, mostly manual but increasingly intellectual.

Effective organization for work is an essential technological capability; so are organizations for education and research. Because technology is so important to production and hence the economy, its structure is intertwined with the structures of industry, business, finance, and markets. Many social institutions, for instance the legal and political systems, treat technology as a minor element. Nevertheless, their decisions on national security, patent law, environmental regulation, and myriad other issues have strong impacts on technological change.

Engineers contribute to knowledge, implements, and organizations through their activities in the science, design, and leadership aspects of engineering. They themselves belong to the fourth and most vital repository of technology, human resources. Technological capacities lie dormant unless they are activated by workers with requisite expertise and skills, from plant operators to business executives. Among the workers, engineers and scientists are peculiar for producing not only goods and services but also more technology. By exercising the technological capability in research, design, and development, scientists and engineers expand it by creating new knowledge, know-how, and implements. In writing papers, teaching, and organizing curricula, they stimulate learning and nurture human resources.

On both the creation and the consumption side, human actions are the fount of technology. Actions are free in principle but constrained in practice. At any geographical and historical point, technological activities are limited by the technological capacity available, economics, politics, and other factors. Activities may excite and subside, but they never cease completely, so long as people exist. Because of incessant activity, technology is dynamic and changes with history. Animated by scientific innovations and differential demands, influenced by luck and cultures, new abilities are continually being created and old ones modified or forgotten. Fluctuations are most complex and unpredictable, but gross currents are discernible.

Phases in the History of Engineering

The history of modern engineering is one of expanding technological capacity. It consists of three overlapping phases. Each has an intellectual aspect for technical contents and a social aspect for human organizations, and each connects to a form of industrial production. After a long prelude, the first phase emerged in the scientific revolution and lasted through the first industrial revolution, when machines, increasingly powered by steam engines, started to replace muscles in most production. While contributing to the revolution, traditional artisans transformed themselves to modern professionals, especially in civil, mining, metallurgical, and mechanical engineering. Practical thinking became scientific in addition to intuitive. Engineering colleges and professional societies emerged.

The second phase brought new disciplines that drove the second industrial revolution, symbolized by electricity and mass production and transportation. Prominent among these disciplines were chemical and electrical engineering, which developed in close collaboration with chemistry and physics and played vital roles in the rise of the chemical, electrical, and telecommunication industries. Marine engineers tamed the peril of ocean exploration. Aeronautic engineers turned the ancient dream of flight into a travel convenience for ordinary people. Control engineers accelerated the pace of automation. Industrial engineers designed and managed mass production and distribution systems. Graduate schools appeared, tinkering became industrial research, and individual inventions were organized into systematic innovations.

Technological advancements in World War II inaugurated the third phase, in which engineering sciences came of age. Astronautic engineering conquered outer space. Use of atomic power brought nuclear engineering. Advanced materials with performance hitherto undreamed of poured out from the laboratories of materials science and engineering. Above all, microelectronics, communications, and computer engineering joined forces to precipitate the information revolution. This period also saw the maturation of graduate engineering education and the rise of large-scale research and development (R&D) organized on the national level.

Besides rapid developments in information technology, new areas for engineering appeared at the turn of the twenty-first century, notably in biotechnology, nanotechnology, and environmental technology. The scope and complexity of these technologies increasingly demand not only interdisciplinary cooperation but also the integration of knowledge in traditional scientific and engineering disciplines. Perhaps a fourth phase is coming that will reconcile the pulls of convergence and specialization.

2.2 Construction Becomes Mathematical

"The constant activity which you Venetians display in your famous arsenal suggests to the studious mind a large field for investigation, especially that part of the work which involves mechanics; for in this department all types of instruments and machines are constantly being constructed by many artisans, among whom there must be some who, partly by inherited experience and partly by their own observations, have become highly expert and clever in explanation."[17] Thus Galileo Galilei opened his *Two New Sciences* in 1638, one of his two major books published in dialogue form.

A key figure in the scientific revolution, Galileo studied not only the heavens but also the productive activities of his fellow citizens, especially in the big land reclamation projects in Venice. He himself held a patent for a water-raising device. More important, his call for systematic explanations of machines, instruments, and other productive artifacts heralded the dawn of engineering science, although it would be a long time yet to sunrise.

Material and Structure: A First Abstraction

Galileo observed that workers in the arsenal made ingenious devices and had many useful explanations for their procedures but were not good at defending their explanations when subject to scrutiny. For example, they knew from experience that one cannot geometrically scale up a small model into a big building. If the model and the building were built of the same material, the former would stand but the latter collapse. Why? Their explanations were vague and were ridiculed by scholars. Defending the artisans, Galileo argued that the same answers could be justified on more reasonable grounds. His own explanations for scaling manifested traits of scientific thinking.

Shipbuilders used more scaffolding per unit area for large ships than for small ships and could estimate the necessary density of the scaffold for a particular size of ship. This was their job. Contented with their success and busy working to make a living, they did not pause to inquire further. Galileo did. He asked *why* and demanded answers that could stand up to scrutiny. He *generalized* his inquiry and regarded the shipbuilding practice as one case in a *type* of process, scaling. Scaling includes many superficially disjointed cases; another example of the concept of scale is the fact that large animals break their legs more easily than small insects when they fall from the same height. Cutting across the diversity of scaling cases, Galileo discerned *material strength* as an important quantity for machines.[18]

Galileo's approach is characteristic of science and engineering. When faced with a particular case, he broadened his scope to seek the *typical* of which the case is an instance. The generalization enabled him to compare various cases, abstract from varying details, discern an important pattern or regularity, represent it in a paradigmatic case, and analyze it into various factors. To study machines, he considered the simplest paradigmatic case, a cantilever beam sticking out of a side of a wall. This he analyzed according to two factors: the material used, which had a certain strength and weight, and the principle of a cantilever supporting a certain load. The analysis brought into relief the material and structural aspects that, after the efforts of numerous workers, flowered into the sciences of *materials* and *statics,* both at the core of civil engineering.

In his analysis, Galileo distinguished what he called "machines in the concrete" from "machines in the abstract." Abstraction, in which an artifact is mostly characterized not in *material* terms but in *structural* and *functional* terms, is a central engineering methodology now known as systems theoretic thinking. Enabling engineers to tackle the material and structural aspects almost separately, it offers great versatility to design. Computer software, for instance, abstracts from most physical details of a computer's hardware, except a few interface parameters such as the amount of random access memory, so that a program can run on any computer that satisfies the parameters. Similarly, the general principles of the cantilever describe machines in the abstract, subject to a few parameters such as the material strength of the lever arm. They are applicable to a wide variety of constructions, such as bridges and tall buildings built with various materials.

Quantification and Measurement of Materials

When stones crack and timbers bend, buildings are in jeopardy. Knowledge of material characteristics is crucial for successful construction. One hundred fifty years before Galileo, master builder Leone Battista Alberti classified woods according to their suitability for various buildings. Alberti's book was the first major work on the architecture and engineering of construction to come off the metal-type printing press introduced in the 1440s. Printing facilitated information dissemination but restricted the *form* of that information. Unfavorable for transmitting hands-on knowledge not expressible on paper, it pushed people to abstract and accentuated the importance of good concepts. The concepts that Alberti used in his classification of materials all pertained to apparent properties, for instance the color or relative durability of types of wood under various conditions. In this he did not surpass his classical predecessors.[19]

Galileo's theory of material strength was a more powerful classificatory concept. Although vitally important for building, material strength was hitherto not a well-defined property, partly because it was so embedded in physical buildings. In contrast to Alberti's qualitative descriptions, Galileo introduced the strength coefficient, a quantitative concept, which was more precise, systematic, and susceptible to measurement. Equally important, the concept of strength is intimately related to that of strain and stress, which are fundamental to structural analysis. These are difficult concepts. Galileo's theories of material strength and the bending of beams had many flaws and errors. It would take more than 150 years—and contributions from Robert Hooke, Leonhard Euler, Daniel Bernoulli, and many others—before Charles Augustin Coulomb and Louis Marie Henri Navier, both French engineers, finally solved the problem of beam bending and flexure. Nevertheless, the first step in the 150-year march had been taken.

Besides concepts, Galileo also introduced the means of *testing* and *measuring*. One can measure material strengths and feed them as parameters into simple although approximate structural equations to get practical answers, without waiting for complicated theories in material science. Soon not only bending and tensile strengths but also the elasticity and crushing resistance of woods, stones, and metals were put to analysis and testing. Many tables of strength coefficients were published. Some were included in Bernard Belidor's *La science des ingenieurs*, which includes perhaps the first use of the term *engineering science*. Published in 1729, the phrase preceded by more than six decades the first appearance of the term *social science*.

Early results on material properties, however, were of limited practicality. They were often unrepeatable and conflicting. Proper comparison and systematization were hampered by the lack of standards for calibration and measurement. The qualities of many materials, such as iron, were unreliable and varied widely with location and manufacturer. Erratic material samples, in turn, undercut attempts to repeat and standardize measurements. As civil engineer and historian Tom Peters remarked: "Reliable materials and standards were, therefore, needed in order to develop reliable materials and standards, a 'Catch 22' situation."[20] Such a situation is the rule rather than the exception in the dynamics of complex processes made up of many loosely linked factors. Each factor moves by both its own internal forces and by external interactions with other factors. Pulling and tugging each other, they constitute the clumsy advancement of the whole field.

Improvements in material quality and reliability came slowly, which partly explains why it was not until 1798 that the first comprehensive modern treatise on strengths of materials was published, by French civil engineer Pierre-Simon Girard. Belonging to the age of stone and timber, the work barely considered iron. In constructing with this new material, France was behind Britain. There an early leader was Thomas Telford, arguably the greatest British civil engineer in the pre-railway era, with numerous roads and canals to his credit. He pioneered the construction of iron bridges. In connection with the proposed Mersey Suspension Bridge in 1814, he used hydraulic machines to perform almost 200 experiments on the tensile strengths of wrought iron chains and cables. He was assisted by mathematician Peter Barlow, who three years later included the data in his own book, which is generally regarded as marking the beginning of materials science in Britain.

Knowledge about materials became more important after 1856, when inexpensive steel was produced in Henry Bessemer's converter and later in Pierre Martin's adaptation of William and Frederick Siemens's open-hearth furnace. A laboratory dedicated to material

studies was established in London in 1865. Material laboratories proliferated as developments in many high-performance technologies came
to depend on the availability of materials with specific properties:
metals, ceramics, polymers, composites. Infused with knowledge from
chemistry and, later, molecular physics, the ancient art of metallurgy
became materials science and engineering, which we will examine more
closely in section 6.4.

Mathematics in Civil and Structural Engineering

In 1742, Pope Benedict XIV commissioned a diagnosis of the cracks in
the dome of St. Peter's Cathedral in Rome. It was a sign of the time that
the commission was given not to artisan-builders but to three mathematicians, one of whom had edited and written commentaries on Newton's *Principia*. Their theoretical approach and conclusion generated a
big controversy. Critics questioned the utility of mathematics in diag-
nosing a dome that was built without using it. Nevertheless, as civil
engineer and historian Hans Straub remarked: "In spite of individually justified objections, the report of the three Roman mathematicians
must be regarded as epoch-making in the history of civil engineering.
Its importance lies in the fact that, contrary to all tradition and routine,
the stability survey of a structure has been based, not on empiric rules
and statical feeling, but on science and research."[21]

The epoch-making event was preceded by a long history in statics,
the analysis of structures in equilibrium. Its basic principles, that of the
lever and the composition of forces, were known since Archimedes and
had been elaborated by Simon Stevin in the sixteenth century. Galileo
introduced the concept of moment and clarified the concept of force.
Statics is a part of the general theory of mechanics, which consumed
most efforts of eighteenth-century physicists and mathematicians. For a
long time, however, theoretical studies had little impact on engineering
practice. Nascent theories tended to oversimplify, failing to capture the
essence of realistic situations. Identifying crucial factors is especially
difficult in engineering design when the structure that will integrate
these factors does not exist yet and design factors can vary in infinitely
many possible ways. It is easier to analyze existing structures, with
which one can proceed like a natural scientist, searching for mathematical representations for given phenomena. Not surprisingly the first en-

gineering application of theoretical statics was to analyze the structure of an existing building, the dome of St. Peter's. The three mathematicians represented the dome's complex construction with a mathematical model simple enough to be analyzed with the laws of mechanics. They calculated forces and concluded that the dome's tie rings were insufficient to contain its horizontal thrust. Their recommendation of adding three iron rings with chains and bolts to secure the building's integrity was adopted.[22]

The design of new structures was more difficult. Engineers familiar with the complexity and demands of realistic situations had to develop the mathematical tools most suitable for the job. Among the pioneers was Coulomb. Trained in the technical academy for French officers at Mézières, he served for nine years in the West Indies as a military engineer before retiring to investigate electromagnetism, in which he discovered the law that now bears his name. Proficient in mathematics, he put it to use in his engineering career to formulate and solve many outstanding problems, including earth pressure, soil slippage, and beam bending, that had taxed Galileo. Unlike mathematicians "with a method in search of a problem," Peters observed, "the value of Coulomb's work lay in the fact he went the opposite way, regarding mathematics simply as a means to an end, using it as a tool, introducing parameters where it suited his intentions and experiences as a builder. Coulomb was not interested in the systematic aspects of the methods he used, but only in their applications. The result was not primarily novel insight or knowledge, but a functioning structure." Coulomb published his work in 1773, writing: "I have tried, as far as I have been able, to render the principles I have used sufficiently clear that an artist with a little learning can understand and use them."[23] Gifted with both engineering and scientific acumen, he bridged the gap between theory and practice, but more than a little casual learning was required for most practical artists to follow. Not until the new technical universities turned out graduates well armed with basic mathematics and scientific principles in the early nineteenth century did theoretical statics come into its own. At the turning point stood Navier, often acknowledged as the founder of structural analysis.

Navier received a sound theoretical and practical education, on the one side from École Polytechnique and on the other from his uncle

Emiland Gauthey, a practicing engineer who rose to the position of inspector general of bridges and roads. Navier designed several bridges across the Seine, inspected and reported on roads and railways in Britain, introduced many theoretical methods for engineering problems, advanced a theory of elasticity, and contributed to hydrodynamics. His biggest contribution in the development of civil engineering, however, was his university lectures, which were first published in 1826. In them he integrated previously isolated discoveries in applied statics and mechanics into a unified theoretical system geared for conceptualizing practical problems and finding solutions for them. Thus he trained his students in that ability most valuable for engineering, in Peters's words, "the ability to base a random case on a system of mathematical principles by intelligent choice of simplifications."[24]

Because of their generality, theories about the real world are always simplifications that leave out many messy and random factors, which scientists can sometimes push aside but engineers cannot. Theories are nevertheless important because they provide insight in the form of conceptual frameworks, by which one can assess new situations, apply accumulated knowledge, and anticipate possible outcomes and alternatives. One difference between a good and a bad theory is its ability to pick the simplifications that capture essential factors that are vortices in the whirl of details and contingencies. Navier synthesized, represented mathematically, and built into his theoretical system some choices important to construction, discovered through his own experience and that of numerous predecessors. Thus he crystallized a body of knowledge related to but different from physics, the engineering knowledge of statics, crucial to buildings of all kinds. Having established its own scientific knowledge base, building and construction transformed from a practical art to a technology.

Mathematics Enters the Field

As Navier lectured, the industrial revolution was at full steam. New building materials—cast iron, wrought iron, and later steel and reinforced concrete—opened design possibilities undreamed of in times of stone and timber. Rapidly expanding railroad networks called for numerous bridges with longer spans and larger load capacities to bear the roaring locomotives. Because these were mostly financed not by governments but by private investors who counted pennies, low costs had to be ensured for both materials and construction processes. Fierce competition further demanded that the structures be designed and erected with unprecedented speed. New supplies, new demands, new problems, and new opportunities sprang up everywhere. The old artisanal way, in which one varied slightly on a narrow stock of tacit knowledge, could neither generate nor keep up with fast and diverse technological changes. Scientific engineering rose to meet the challenge —a theme that recurs in modern history in cases of increasing scientific and technological complexity.

For a small example in the big progress of civil engineering, consider the construction of the Britannia Bridge across Menai Strait. The bridge was part of the railway line linking London and the terminal in Wales for the ferry to Dublin. Robert Stephenson, the line's chief engineer, started working on the project in 1845. He soon decided that the suspension bridge that had been built by Telford twenty-five years earlier was too flexible for locomotives. A new bridge with a stiffened deck was necessary. He enlisted William Fairbairn, a shipbuilder with twenty years of experience in wrought iron construction, and Eaton Hodgkinson, an engineer dubbed "mathematician" by his colleagues because of his capacity for theoretical analysis. The very choice of personnel reflected a change in time and mentality. When Telford planned his bridge, he suspended a full-size chain to measure its sag, instead of calculating the sag with a catenary equation. In contrast, Fairbairn routinely used the science of hydraulics to determine hull shapes, and he called on Hodgkinson's mathematical analysis for the bridge.[25]

Stephenson's team decided on a hollow-box beam bridge. A beam 1,380 feet long sitting on three intermediary towers was a daring novelty, but they were not totally in the dark. Navier had analyzed the mechanics of continuous beams with multiple supports, and his work was introduced in Britain by Henry Moseley in 1843. The mathematics involved was far from sufficient, not the least because its reliability was questionable owing to a lack of thorough testing. It did not touch on problems peculiar to tubular beams, such as buckling, and was too crude to yield detailed specifications. The design process was mainly empirical, and quantitative decisions were based on experiments. However, the theoretical framework did help in deciding what experiments to perform, what data to take, how to analyze them and draw their significance for the proposed bridge, and what follow-up tests to pursue.

For construction of the Britannia Bridge, many of Hodgkinson's theoretically motivated experiments were blamed for wasting time. For civil engineering, however, they were anything but a waste. They revealed problems, discovered solutions, introduced ideas, tested theories, and advanced construction technology. Maxwell observed that the application of electromagnetic theory to telegraphy helped to advance the science itself. Similarly, the application of structural analysis to bridge building reacted on the science of statics by giving a commercial value to accurate mechanical measurements, and by affording to engineers the use of apparatus on a scale that greatly transcended that in laboratories. By assimilating and generalizing data gleaned from the bridge's construction, mathematical theories acquired more substance and credibility—and they would be more useful for subsequent construction projects.

Architecture and Engineering Diverge and Reapproach

Construction, which accounted for more than 4 percent of the gross domestic product of the United States in 2000, is a large industry. Roughly 15 percent of it is heavy construction, which includes highways, streets, tunnels, bridges, dams, levies, landfills, water and other pipelines, power and communication transmission lines, and industrial factories.[26] This is the empire of civil and structural engineers. An exceptionally large percentage of them work for the government, because many civil infrastructures are public undertakings.

Another 45 percent of the industry is involved in constructing buildings. Here lies the story of the architectural and engineering professions. The industrial revolution ushered in a new construction method the potential of which has not yet been exhausted today, the structural use of iron and steel. Building high towers, or roofs over huge spaces, requires structural analysis and accurate measurement of material strengths, the expertise of civil and structural engineers. Charged with building train stations, exhibition halls, factories, and other facilities that emphasized function, engineers started to build without consulting architects. The two professions split and were for a while estranged. The gulf between them was most apparent in the contrast between

London's St. Pancras Hotel and the adjoining train station. The hotel was an ornate Gothic structure and the station a plain vault of iron and glass, with no sign that builders had tried to mitigate the disharmony. This does not mean that the utilitarian style is devoid of beauty. The station's sweeping spaciousness has a breathtaking sublimity. The aesthetic problem lies in its clash with the adjoining hotel. For the caption to a picture of the station, historian of architecture Furneaux Jordan wrote: "The iron roof by W. H. Barlow which spans the 243 feet of St. Pancras Station in London (1864) is one of the finest engineering achievements of the 19th century. Note how the great curve of the girders dwarfs the little Gothic windows of the hotel building, joined on at the end."[27]

Ornament eventually gave way to technology. Historian Henry Hitchcock observed: "Historians of modern architecture have generally emphasized, and rightly, the special importance of the advances in metal construction that were made in France in the later decades of the nineteenth century. The great name of the period is not that of an architect but of an engineer, Gustave Eiffel."[28] Eiffel built several magnificent bridges, but most people know him by the tower that he built in 1889 in Paris. Soaring upward for 300 meters in naked structural iron, the Eiffel Tower had not only the physical strength to withstand the elements but also the social strength to withstand the mountain of scorn heaped on it by aesthetes and eventually become one of the most beloved landmarks on earth.

As an engineering structure, the Eiffel Tower is exceptional for its conspicuousness. Aside from bridge towers, which stand exposed, the engineer's towers usually share the fate of another of Eiffel's works, the internal structure supporting the sculpted skin of the Statue of Liberty, invisible to and easily overlooked by all except those who take the tour inside. The Empire State Building, which broke the height record held by the Eiffel Tower for forty-two years, hides a structural steel frame. This type of skyscraping architecture was pioneered by another graduate from the same engineering college that produced Eiffel.

After studying in France, William Le Baron Jenny returned to his native America and served for seven years as an army engineer before turning to the practice of architecture. He became a leader of the Chicago School, and many of its members passed through and learned from his firm. The Council on Tall Buildings and Urban Habitat states: "Jenny's Home Insurance building of 1884–1885 initiated the innovative use of the structural steel frame which would characterize much of subsequent tall building design."[29] Previous technology for structural masonry had been limited to load-bearing walls, which led to extremely thick walls for high-rises, culminating in the six-feet-thick base walls for the sixteen-story Monadnock Building. Home Insurance was innovative in that almost all of its exterior walls were not self-supporting but were carried by an internal steel frame that bore the building's weight. Although only ten stories high itself, it introduced the novel construction technology that made the sky the limit for modern buildings.

Unlike small buildings whose structures impose relatively few physical demands and thus leave room for arbitrary variations in form, a skyscraper's structure is so physically demanding it must be thoroughly integrated with the building's form. Architects dominate the design of family houses, but for tall buildings they must work closely with structural engineers. In thinking about a building, architects emphasize the human factors, its aesthetics and service to life and business; engineers emphasize the physical factors, its structural integrity in contending with natural forces. Architects can concentrate only on the building as a finished product, how it looks and functions. Engineers cannot; they must reckon with not only the design of a building but also the construction process that implements the design and turns it into physical reality. Complementing each other's expertise, architects and engineers often team up to form a single firm offering packaged services. The two professions that originally split from the ancient *architectus* eventually reapproach and collaborate.

2.3 Experimenting with Machines

James Watt's steam engine made its commercial debut in the same year that the shot heard round the world was fired in America. While the French Revolution of 1789 raised the vision of liberty, equality, and fraternity, Britain raised that of prosperity, as increases in its rates of production and trade jumped. This was the period known as the industrial revolution, an era that propelled Britain into the lead among the world's economic and technological powers.

Those who made good Britain's claim of being "the workshop of the world" were pioneer engineers, especially in mechanical engineering, who gave the world locomotives and other symbols of the industrial revolution. Watt is sometimes called the father of mechanical engineering. His career exhibited the complexity of the engineer, because he combined attributes of the scientist, practical artist, and entrepreneur. He performed scientific experiments and tried to grasp some principles of engine operation. He excelled in designing all kinds of devices and instruments besides the steam engine: the flyball governor, the pressure gauge, the stroke counter, the sun and planet gear that translates piston to rotary motion, and more. Besides inventing and improving machines, he also formed a partnership to produce and market them, so that they actually served society and enriched him.[30]

To understand engineering as exemplified in Watt, recall the two aspects of technology, the organizational and the physical. Organizational technology, which concerns the orchestration of human and material resources for production, includes entrepreneurship and business administration. Many mechanical engineers, including Watt, Fairbairn, and Stephenson, were also entrepreneurs. Physical technology involves "physical" forms in the widest sense, including computers and information processing algorithms. Four basic platforms of physical technology that enable higher-level constructions are *materials, power, tools,* and *control.* We touched on materials in the preceding section and will look at the other three platforms here. Steam engines provided power that freed production from muscle and geographical locations of wind and water. The engines themselves required adequate machine tools for their production and control mechanisms for operation. Power, tools, and control were first developed in mechanical engineering, although other engineering branches joined in as physical technology advanced.

Scientific Empiricism in Engineering

Civil and mechanical engineering differ in the general characteristics of their products and their processes of production. Civil engineers produce constructions, mechanical engineers machines. Constructions are rather homogeneous in function and form. Although each suspension bridge poses problems peculiar to its requirements and environment, it shares salient features with other suspension bridges. Thus civil engineering structures are more susceptible to generalization and mathematical representation, and structural analysis was the first engineering science to mature. In contrast, machines serve countless complex functions. Their forms and operations, diverse and specialized to their functions, are less prone to generalization and theorization. General principles of heat or fluid flow that underlie various types of engines were discovered later and only with significant input from engineering practices. Mechanical engineering has tended to be more empirically oriented, especially in its early days when the relevant sciences were embryonic. Natural science was not irrelevant, but its contributions were more in general ways of systematic reasoning and controlled experimentation than in specific theories and pieces of information. This is illustrated in Watt's experience.

It is a myth that Watt got the theory of latent heat from Joseph Black and applied it to invent a separate condenser for his steam engine. Watt independently discovered the effects of latent heat in his own experiments. It is important to note that his experiments were not the tinkering of a traditional artisan. He had worked as an instrument maker for a university laboratory, and this experience made a big difference. He wrote: "Although Dr. Black's theory of latent heat did not suggest my improvements on the steam engine, yet the knowledge upon various subjects which he was pleased to communicate to me, and the correct modes of reasoning, and of making experiments of which he set me the example, certainly conducted very much to facilitate the progress of my inventions."[31]

Reasoning and experimentation, which Watt emphasized, are characteristics of scientific empiricism that separate engineers from craftsmen. Watt was introduced to the steam engine when asked to fix a model used for class demonstrations. After fixing the little engine, he played with it and noticed that it ran out of steam quickly. He wondered why the big industrial engines conserved steam so much better. To find out what spoiled the linear scaling, he analyzed the engine, examined its components individually, and identified the culprit as the difference in areas of condensation surfaces. To solve the problem, he measured the volumes and temperatures of the steam and the condensed water, discovered the effect of latent heat of condensation, determined to find ways to prevent it from harming engine efficiency, and came up with the novel idea of a separate condenser. This is neither blind "cut and try" nor tinkering but reasoned inquiry and creativity, characteristic of scientific empiricism.

The nature of scientific empiricism in engineering is revealed more clearly in the experiments of John Smeaton, the first Englishman to call himself a civil engineer, in contrast to a military engineer. He performed many experiments, one of which aimed to determine the relative efficiencies of various types of waterwheels. If Watt's steam engine was a revolutionary breakthrough, Smeaton's waterwheels exemplified an evolutionary improvement, which is no less important for the advancement of technology.

As water power was harnessed for industrial purposes, waterwheels increased in size and complexity. Two types were widely used. With an *overshot wheel,* water comes in from the top, fills buckets attached to the wheel's periphery, and turns it by the force of gravity. With an *undershot wheel,* water in a stream rushes past the wheel's paddles and turns the wheel by impulse. A theory existed for the superiority of undershot wheels. Finding that the theory deviated from practical experience, Smeaton conducted a series of experiments in 1752. He built a model wheel two feet in diameter, variously equipped at the periphery with buckets or paddles. It was attached to a variable load. The reservoir that supplied its water was designed with controllable head and flow rate. Thus the experimental setup involved four interrelated parameters: load, head, flow, and wheel periphery. He varied the value of one parameter, keeping the other three constant, and for each setting measured work output as the product of the load and the wheel's turning speed. His data overthrew theoretical results, determined optimal operating conditions, and suggested improved designs that could double a wheel's output.

Smeaton's work is also valuable for demonstrating a general empirical approach called the *controlled experiment* in natural science and *parameter variation* in engineering. Unlike demonstration models in the feasibility studies of particular designs, Smeaton's model water-wheel is a scientific study of a *general type* of design and its underlying mechanisms. Like natural scientists who create a vacuum or other phenomena in the controllable environment of a laboratory, Smeaton created his experimental apparatus. He distinguished relevant parameters, varied them systematically, teased out their individual effects and mutual relations, and offered reasoned interpretations of his data. These are hallmarks of the analytic approach in natural scientific experiments. From his data, he drew certain conclusions about the primary physical causes for the working of waterwheels. He published these results in a scientific paper. Robert Stephenson, who came almost a century later, wrote: "To this day there are no writings so valuable . . . in the highest walks of scientific engineering; and when young men asked me what they should read, I invariably said: 'Go to Smeaton's philosophical papers; read them, master them thoroughly, and nothing will be of greater service to you." At that time waterwheels were becoming obsolete, but Smeaton's thinking and approach were not.[32]

The importance of controlled experimentation does not imply that it alone is sufficient for scientific progress. Smeaton's conceptual framework was rather crude. It sufficed for simple machines such as water-wheels, but as machines became more complicated, they called for more sophisticated theories that may be less efficient for simple cases. Working at the same time as Smeaton but independently from him, mathematician Johann Euler used his father Leonhard's fluid theory to analyze waterwheels and arrived at similar conclusions about the superiority of gravity wheels. His work won him a scientific prize but was largely ignored by millwrights because it was opaque to them and offered no more practical advice than Smeaton's empirical results. However, his mathematical theory suggested a more general type of hydraulic machine and formed an important step in the subsequent development of water turbines, which now hum in hydroelectric plants while old-fashioned waterwheels have disappeared.

Power from Energy Conversion Engines
Waterwheels, windmills, steam engines, internal combustion engines, and their like constitute a class of machines called *prime movers*, or *en-*

ergy conversion engines. They are the heartbeat of industry, because they unlock the energy in fuel, water, and other primary sources and turn it to useful power or mechanical work. Their role did not diminish with the rise of electricity, which is a versatile intermediary whose generation depends on prime movers in power plants to extract energy from natural sources.[33]

Designing energy conversion engines is a major topic in mechanical engineering. As engines become more complex and sophisticated, their designs increasingly require adequate understanding of underlying physical principles, to which both engineers and physicists contribute. An engineering history that followed the line of energy conversion would trace the evolution of various types of power-generating engines: how waterwheels became water turbines; how steam engines powered the railway age and eventually evolved into the steam turbines that work in fire-power plants and large ships; how internal combustion engines conquered the road and the sky, both in their reciprocating versions as in propeller aircraft and their turbo versions as in jets. We would see how these engines, when operated in reverse, become pumps and refrigerators. We would note the convergence of various types of engines toward turbine designs. We would marvel at the ingenuity with which engineers manipulate compressible fluids as media to convert heat from burning fuel into useful mechanical work.

Advancement in engine design is accompanied by advancement in scientific understanding of underlying mechanisms. Along this line we would see how, in studying the operations of the steam engine, French engineer Sadi Carnot initiated thermodynamics; how the Carnot cycle and science of heat in turn facilitated the design of other heat engines, especially the efficient diesel engine that powers most commercial vehicles. We would examine Charles Parson's reaction steam turbine; ask how designs of efficient turbines demanded scientific understanding of fluid motion and compression; how they stimulated developments in hydrodynamics and aerodynamics; how scientific theories, in turn, made possible the invention of gas turbines for jet propulsion.

The history of energy conversion illustrates the nature and role of engineering sciences such as thermodynamics and fluid dynamics, which place natural phenomena in productive contexts and systematically explore their general principles under practical constraints. It offers another example of the coupled dynamics of scientific theory and engineering design. For lack of space, I will pass it here; we have explored this theme in the case of civil engineering and will encounter it again in other areas.

From Steam Power to Automatic Machines

Watt's innovative engine design sat on the shelf for five years because no one had the skill and tools to make it. It would have sat there longer had John Wilkinson not come up with a boring machine capable of making the engine's large-bore cylinder. Energy conversion engines drove the industrial revolution. Being complicated machines, they themselves could not get off the ground without adequate *machine tools,* the machines that make machines.[34]

The bow drill appeared in Egypt at about 2500 B.C.E., and evidence for lathe work in Etruria dates back to 700 B.C.E. When Leonardo sketched lathes, grinders, boring mills, and screw cutters in his notebooks, similar machine tools were already in use and continually being refined. Early machine tools were for woodworking, and wood was increasingly replaced by metal as industrialization progressed. Machine tools that make modern machines are mostly metalworking devices, capable of accurately cutting or forming large quantities of metallic parts, big and small. Metal is much more difficult to work with than wood, and machine tooling requires much engineering.

Manufacturers of steam engines, locomotives, and steamships played double roles in the machine tools industry. Besides demand pull, they exerted technological push by operating their own machine shops. When Wilkinson quit business in 1795, the company of Boulton & Watt set up its own foundry, complete with shops for drilling, heavy turning, gear cutting, patterning, fitting, and other jobs. Its archrival, Matthew Murray, promptly followed suit. These two industrial engineering shops, the first of their kind at least in the private sector, competed in developing new techniques for making engine parts, such as a planing machine for making the flat surfaces of slide valves. They became schools for numerous mechanical engineers, many of whom subsequently founded their own shops. As it grew, the machine tools industry attracted many talents. Among them was Henry Maudslay, who pioneered precision machine tooling. His machine shop nursed many mechanical engineers, for example James Nasmyth, who invented the steam hammer, and Joseph Whitworth, who brought precision to a new height.

One factor behind Britain's prowess during the industrial revolution was the superior machine tools developed by its pioneering engineers, who in fifty years turned the age of steam power into the age of machines. The transformation of the technological landscape was described by Fairbairn, who came to Manchester in 1813 and addressed a city assembly forty-eight years later: "When I first entered this city the whole of the machinery was executed by hand. There were neither planing, slotting nor shaping machines; and with the exception of very imperfect lathes and a few drills, the preparatory operations of construction were effected entirely by the hands of the workmen. Now, everything is done by machine tools with a degree of accuracy which the unaided hand could never accomplish. The automaton or self-acting machine tool has within itself an almost creative power; in fact, so great is its power of adaptation that there is no operation of the human hand that it does not imitate."[35]

Even as Fairbairn spoke, the leadership in machine tools was being wrested from Britain by the United States, which would retain it through the 1970s. Americans invented two powerful general-purpose machine tools, the milling machine and the turret lathe. In the 1950s MIT, supported by the aircraft industry and the U.S. Air Force, developed numerical control, which brought automation of machine tools to a new level. Today more than three-fourths of the machine tools sold in America are computer controlled for precision and flexibility.

"The machine-tool industry stands at the heart of the nation's manufacturing infrastructure, and it is far more important than its relatively small size might suggest," wrote the MIT Commission on Industrial Productivity, which worried about the industry's decline in America in the 1980s. All industries depend, directly or indirectly, on machine tools to cut and shape metal parts. They risk losing their competitive edge if their tools are inferior in speed, precision, reliability, flexibility, or cost-effectiveness. When Detroit auto manufacturers complained that the best and fastest machine tools they could buy had already been in use at BMW or Toyota for two years, trouble was brewing for the American automobile industry.[36]

Physical Embodiment of Knowledge in Machines

As machines proliferated, their working mechanisms were systematically studied. Franz Reuleaux's *Kinematics of Machines* of 1875 analyzed complicated machines, explained how machine parts generally moved relative to each other, and delineated general chains of mechanism. Such conceptual generalization and discursive exposition constitute what agriculture engineer Born Douthwaite calls "disembodied knowledge." Valuable and indispensable as it is, it does not exhaust knowledge. People learn from experience that often no verbal description or even picture can substitute for seeing the real thing as it works. That is why information technology does not eliminate the necessity of field visits. Douthwaite explains how the concrete machines themselves, especially automatic machines that perform tasks that previously required human workers, constitute "embodied knowledge." He uses agricultural machines as examples; machines tools can do as well.[37]

The limited number of basic processes for shaping metal fall into two major classes. Metal forming processes include casting, stamping, punching, rolling, and forging. Metal cutting processes include turning, milling, grinding, drilling, and boring. Each process involves several tasks that, although rather simple individually, must cooperate in complicated ways for making intricate pieces. Craftsmen have invented numerous hand tools for individual tasks. Their skills reside in using the tools in concert to produce desired results. Those *skills* are what engineers try to automate. Among myriad ways in which workers perform a process, engineers try to discern a few important characteristics and physically relate them in an integral machine design so that when those characteristics are varied, the design adapts to provide first-rate performance in a wide range of conditions. Engineers distill essential characteristics of a type of skill and build them into a type of general-purpose machine tool. Thus automation effects a *physical generalization* that embodies know-how in concrete machines, in contradistinction to *conceptual generalization* that represents knowledge in words or equations.

In contrast to the analytic tendency of conceptual explanations, physical embodiment of knowledge emphasizes the synthesis of various components into a complex functioning whole. Consider, for example, the lathe, one of the oldest machine tools. By 1800, all components of a modern lathe—spindle bearings, lead screws, slide rests—were in use somewhere separately. Then Maudslay modified them for integration in a rational design. He took pains to ensure that they worked together with great flexibility, reliability, and mechanical controllability, thus making the lathe a basic tool for precision machining. Synthesis is often deemed the core of engineering. Mechanical engineer and historian R. S. Woodbury wrote that Maudslay "provided the great synthesis that embodied all these earlier elements in a design that set the fundamental form which the lathe was to have down to the present, the form which was to make the lathe a profoundly significant tool."[38]

Another example is Joseph Brown's universal milling machine, which distilled and synthesized insights from almost half a century of milling practice. Although originally developed for making grooves in twisted drills, the milling machine proved to be immensely versatile. After mastering the milling machine in making grooves, a worker can readily modify it to suit jobs in other industries. Like Maudslay's lathe, Brown's milling machine became a basic, general-purpose machine design that has persisted through endless adaptations for new applications. These basic machine designs are like scientific principles, except they embody general knowledge in concrete rather than abstract form.

Automatic and Feedback Control

Nothing embodies skill and know-how in physical form more effectively than control devices. A craftsman controls the motion of a hand tool and its position and angle relative to the work. The hand tool itself has no control. As technology advances, more and more of the worker's control is being incorporated into the machine tool. Maudslay's screw-cutting lathe with slide rest, for instance, achieved significant control by mechanically integrating translational motions of the cutting tool with respect to the rotation of the blank. Much more is to come.[39]

A general-purpose machine must allow for considerable adjustment and leeway of motion so that it can be used to fabricate a wide range of shapes. Without additional equipment, the adjustment is controlled by a human machinist. To make the machine work automatically in fabricating a particular shape, engineers add a controller to it. Nowadays,

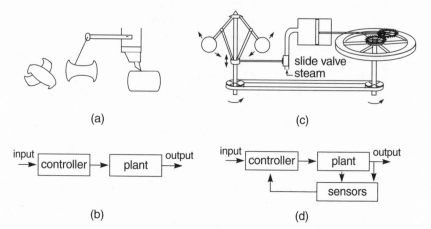

Figure 2.3   (a) A template controls the movement of a machine tool to produce items of identical shape. (b) Abstraction from the machine tool reveals an open-loop control system. (c) Watt's flyball governor consists of two flyballs swung by a rotating shaft that is connected by belts to the shaft of the rotary steam engine. When the engine accelerates undesirably, the acceleration makes the flyballs swing faster and lift up because of centrifugal force. The lifting moves a sliding valve, closes it proportionally, reduces the steam to the engine, and thus slows it down. (d) Abstraction from the self-regulating engine yields a feedback control system.

most controllers are digitally implemented algorithms. In the old days they were jigs, fixtures, and limiting switches, which work on the same basic principle as a key-duplicating machine. A stylus moving on a template guides the motion of the machine tool, which is stopped by a switch at the desired position, as illustrated in Fig. 2.3a. Although expensive to create, jigs and fixtures are cost-effective for manufacturing many pieces of the same shape. They were crucial to the rise of mass production.

In control engineering, the system to be controlled is generally called a *plant*. Control schemes fall into two large classes. In *open-loop* control, illustrated in Fig. 2.3b, the controller contains all information necessary to adjust the plant's behavior through an actuator without further input. In more sophisticated schemes, a sensor measures the plant's state, and the controller uses the measured information in deciding what adjustment to make to the plant. This is called *feedback* control, as illustrated in Fig. 2.3d. Whereas an open-loop controller merely issues commands, a feedback controller constantly monitors the motion of the actuator and the response of the plant and compensates for any execution error in its subsequent commands.

The term *feedback* was coined in the 1920s by communication engineers. However, closed-loop control mechanisms have been in use since antiquity, for example in the oil lamp made by Philon of Byzantium in the third century B.C.E. Inspired by the mechanism used to regulate the spacing between grinding stones in water mills, Watt designed the flyball governor in 1788. Coupled to the steam engine, the two form a self-regulating system that operates at a constant speed (Fig. 2.3c). The flyball governor belongs to a class of feedback control called *regulators,* which aim for a plant's stable operation by maintaining one of the plant's variables at a constant value. A familiar regulator is a thermostat that maintains constant temperature. Besides in machines, regulators are used in industrial plants to regulate temperature, pressure, flow of fluid or gas, electrical current and voltage, and other variables under diverse operating conditions. Because regulators must work in conjunction with sensors that measure the values of the relevant variables, they stimulate demand for measuring instruments. Engineers respond by inventing a plethora of mechanical, electrical, pneumatic, and other kinds of sensors and transmitters of information.

Great steamships brought the need for another kind of controller. Not even Hercules could turn the rudder of a ship that has thousands of tons in displacement. To move the rudder, the manual force applied at the helm must be amplified, probably by a steam engine. Engineers had to find ways to make the rudder respond precisely to the movement of the helm and give the helmsman some "feel" about the rudder's behavior. In 1866 McFarlane Gray patented a steering engine incorporating feedback control that found first employment in the 21,000-ton *Great Eastern.* It represented a new type of controller that modified the value of a plant variable to track precisely the changing value of an input variable: the motion of the rudder tracking the motion of the helm. This new controller is called a *servomechanism,* from the Latin *servo,* meaning slave or servant.

Flyball governors sometimes go into oscillation. Similar behaviors plague many other kinds of control systems, which are often unstable or easily destabilized by small changes in system parameters. These problems attracted the attention of physicists and mathematicians who had a traditional interest in the stability of dynamic systems, especially the solar system. In 1868 Maxwell studied governors as dynamic systems represented by differential equations, and eleven years later Edward Routh found a stability criterion. As applications multiplied and system complexity soared, problems also multiplied whose solutions demanded mathematical analysis. Soon engineers such as Nicolas Minorsky and Harold Hazen took up theories for automatic control. Minorsky analyzed performances of various types of steering servomechanisms in 1922. He carefully studied the performance of human helmsmen, whose skill and knowledge he tried to incorporate in controllers, not only physically but also theoretically. Discovering that the rudder's deviation from its desired position is as important as the rate change of its deviation, he arrived at a clear formulation of the three-term PID (proportional, integral, derivative) controller that corrects transient, steady-state, and destabilizing errors. A similar steering mechanism had been constructed by Elmer Sperry eleven years earlier relying on rational empiricism, but Minorsky's theoretical analysis opened up vast new territory. Partly because of its conceptual clarification, the PID controller has been generalized to many applications and is still in wide use today.

Controlling the operations of machines and plants is a common aim for all branches of engineering. When mechanical engineers struggled with servomechanisms, electrical engineers wrestled with stabilities of electricity transmission systems and amplifiers for long-distance telephony. Brought together during World War II, engineers discovered that certain general principles underlay apparently diverse control processes in various fields. Cross-fertilization bore fruit after the war, when engineering surged forward on its third wave of progress. Among its achievements were sophisticated control theories, to which we return in section 6.2.

2.4 Science and Chemical Industries

What are the most salient characteristics of modern living as we know it today? Probably electricity, cars, mass production, and mass consumption would sit high on the list. These are the fruits of the second industrial revolution, which occurred in the late nineteenth and early twentieth centuries.[40] Here we concentrate on the electrical and chemical industries, which have a great impact on modern society's infrastructure. Science-intensive from their inception, they were the businesses that inaugurated industrial research. Spearheading the new industries were new groups of engineers. Aware of growing opportunities to serve their own careers as well as society's productive needs, students flocked to engineering courses, especially chemical and electrical engineering.

Unlike civil and mechanical engineering, which have deep roots tracing back to antiquity, chemical and electrical engineering have little tradition except a staunchly practical attitude. They stem from chemistry and physics and are imbued with the thinking of those sciences. Pioneers of chemical engineering were mostly university professors, who emphasized the importance of science and basic principles. After an initial wave of self-taught engineer-entrepreneurs, leadership in American electrical engineering passed to men such as Charles Steinmetz and Michael Pupin, who, versed in mathematics and physics, set the profession on the path of rigorous reasoning. By the 1930s, electrical engineer Dogald Jackson observed: "The precise measurements and controlled experiments introduced into our field from the field of physics gave a tremendous impetus to rational and accurate engineering calculations and also left a scientific impress on electrical engineering teaching."[41] Electrical and chemical engineers became leaders in the development of engineering sciences.

Chemistry, electromagnetism, and later quantum mechanics unlocked the doors to previously arcane physical phenomena. The new areas opened up are enormous and complex. To explore and exploit them, scientists and engineers have joined hands. Yet chemical and electrical engineering are no more appendixes to chemistry and physics than a man is an appendix to his sibling. They have different orientations and develop different although sometimes overlapping substantive contents. In pharmaceuticals, for instance, chemists tend to work more on drug discovery and chemical engineers on efficient processes for drug manufacturing. In telecommunications, physicists tend to work more on devices and electrical engineers on networks. Chemical engineering naturally extends into biochemical processes and biotechnology. Electrical engineering has expanded to cover microelectronics and computer science, and with its command of telecommunication systems it is on the leading edge of information technology.

From Chemistry to Chemical Engineering

Since Antoine Lavoisier's work in the 1780s, chemistry broke with its alchemical past and became a science to be pursued for its own sake, which eventually led to atomic science. Nevertheless, the practical uses of chemicals in medicine and other areas were always appreciated and pursued. Practicality enhanced the popularity of chemistry. Starting in the late eighteenth century, new impetus came from the blossoming textile industry. Cloth bleaching and dying urgently required chemical replacements for vegetable alkali, which was in short supply. The demand was met by the Roebuck lead chamber process for sulphuric acid and the Leblanc and the superior Solvay processes for caustic soda and bleaching powder. Britain led in the production of these heavy chemicals, but by the end of the nineteenth century, it was being passed by the United States.[42]

Organic chemistry emerged in 1810. After William Perkin discovered the first aniline dye in 1856, chemists all around Europe raced to react aniline with every chemical on their shelves to create a rainbow of colors. After August Kekulé's 1865 discovery of benzene's hexagonal structure, chemists were aided by theories in their search for organic compounds useful as dyes or drugs. The BASF, Höchst, and Bayer corporations were formed around 1863 by chemists and people with technical knowledge of chemicals, who appreciated the importance of scientific knowledge in their industry. These German chemical firms established close relations with universities and developed industrial research laboratories that became models for the rest of the world. Research enabled them to invent hundreds of dyes and produce them at lower unit costs than their competitors, develop novel products and processes, and diversify into other areas such as pharmaceuticals. Germany quickly overtook Britain in fine chemicals and went on to dominate in synthetic dyes.

Germany was the world leader in chemistry and the chemical industry, but it was not the Germans who developed chemical engineering as a discipline with its distinctive body of scientific concepts and knowledge. Chemical engineering was envisioned by George Davis in Britain and brought to fruition by Americans, MIT's William Walker and Warren Lewis and their close associate, engineering consultant Arthur Little. All three, like many of their contemporaries, had studied in Germany. In 1888, MIT offered the first four-year curriculum labeled chemical engineering, which became an independent department in 1920. The University of Pennsylvania was the second, in 1892. Most chemical engineering departments originated in university chemistry departments. Despite their name, early contents of their curricula were mainly chemistry plus mechanical engineering.

Contents distinctive of chemical engineering were developed in the early 1900s. Walker, Lewis, and Little delineated and systematized unit operations, the basic processes general to many types of chemical reactors that engineers can mix and match in reactor designs. With this core curriculum, chemical engineering attracted more and more students.

In 1908, Davis's vision was realized with the establishment of the American Institute of Chemical Engineering, but not without considerable difficulties from the American Chemical Society. While electrical engineering rapidly established its identity distinctive from physics, chemical engineering and chemistry have a stronger bond, or what chemical engineer L. E. Scriven called "a sometimes love-hate relationship."[43] Chemistry is more inclined toward application and commerce than is physics, a difference one notices today when comparing the American Chemical Society's *Chemical & Engineering News* and the American Physical Society's *Physics Today*. To distinguish itself from chemistry, chemical engineering had to have more to show than practicality. By integrating chemistry, physics, and mathematics, chemical engineers created an engineering science that bears on a wide class of industrial chemical processes. With its scientific ability, it has built not only a flourishing profession but also many new industries based on new processes.

Products and Processes

Germany led in chemistry and chemical industry. Why did it not develop chemical engineering? Here we come to the important distinction between a *product* and its *process of production*. Products do not materialize magically from designs on paper. Sometimes even the simplest of products demand sophisticated processes, especially when cost, quantity, and time are at issue. Aluminum is a mere element and the most abundant metal on earth, but it took a long time to develop the electrolytic process for producing aluminum from bauxite on a commercial scale. It is for brevity only that I focus on products such as bridges and steam engines when discussing civil and mechanical engineering. Processes of construction and manufacturing are equally important and often impose constraints on product designs. Stephenson had to forgo his idea of an arch for the Britannia Bridge because its construction process required temporarily blocking the strait, which the British admiralty forbade. Because such ordinances are more the rule than the exception, it is almost mandatory for modern bridges that they be self-supporting at each stage during construction. Constraints on a bridge's building process strongly influence its structural design. *Design for manufacturing,* a motto of product engineers, indicates the importance of taking production processes into account.[44]

Designs of product and process demand comparable intellectual ability. This is epitomized by the two chemistry Nobel Prizes awarded for the Haber-Bosch process. Fritz Haber received the 1919 prize for discovering a way to synthesize ammonia. Carl Bosch shared the 1931 prize for developing the process to manufacture synthetic ammonia. Brilliant as it was, the chemist's discovery would have remained in the laboratory if the engineer had not overcome enormous difficulties to achieve the large-scale production that made synthetic fertilizer beneficial to the mass of mankind. It is in production processes that chemical engineering finds it niche.

Industrial chemical processes are not simply larger versions of laboratory chemical reactions. Besides constraints imposed by cost, production volume, and environmental impact, many hurdles obstruct scaling up chemical reactions from the test tube to the industrial tank. A chemist can shake a flask over a flame to heat and mix up the reactants for a desired reaction, but a similar method applied to a thousand-gallon tank could end in a deadly explosion. Large containers, with their smaller surface-to-volume ratio, are more inimical to heat distribution. To heat their contents, close attention must be paid to processes of fluid motion and heat transfer to ensure the proper heating and mixture required for the chemical reaction. The physical processes relevant to industrial chemical reactors can be systematically discovered and their general principles studied. Then they can be readily used to scale up a wide variety of chemical reactions for efficient industrial production. It is in these tasks that chemical engineering distinguishes itself from chemistry.

Processes are more important in the chemical industry than in other industries for both technical and economic reasons. Technically, the connection between products and processes is especially strong for chemicals. Many chemical products, such as polymers, have no definitive blueprint. Although their molecular structures are specified, their final properties vary according to how they are produced. The contribution of production processes to product properties is even greater in advanced and specialty materials that will be increasingly important in the future. For such cases, the early involvement of process engineers is necessary to guarantee satisfactory product characteristics.

Economically, the chemical industry is more intensive in capital and raw materials than in labor. Many chemicals are industrial commodities that are consumed in large quantities. For high-volume production, the cost of chemical plants and other capital investments can consume up to 50 percent of the sales price of a product. Few can afford to build such expensive plants and then experiment with them to figure out adequate production processes. Thus the processes must be understood to a significant extent first, to predict plant behavior even in the planning stage.

Chemistry is more involved in the design of products—chemical molecules and reactions; chemical engineering, in the design of efficient manufacturing processes. In the nascent chemical industry, the job of scaling up chemical reactions from laboratories to industrial production was accomplished by chemists teaming up with mechanical engineers. This interdisciplinary cooperation was satisfactory to the Germans, partly because they excelled in sophisticated products: organic and fine chemicals, such as dyes and pharmaceuticals. These complicated, specialized, and high-value products required sophisticated chemistry but not much engineering. Scaling up was not demanding because their production volumes were small, and their high market prices could cover relatively high manufacturing costs. In 1913 Germany produced 137,000 tons of dyes, consisting of thousands of different dyes, while the United States produced 2,250,000 tons of sulphuric acid alone. The American and British industries concentrated on the mass production of commodity chemicals, such as acids and soda. These simple and cheap products required little chemistry but much engineering to scale up production to huge volumes. Their razor-thin profit margins put the spur to cutting production costs. It was no surprise that the Anglo-Americans were keener than the Germans to develop the science and technology of large-scale production processes—chemical engineering.

Unit Operations

Many chemical processes were already industrially employed in the nineteenth century. To account for them a branch of chemistry was developed. Industrial chemists treated a production process as a unit without analyzing it into components and generalizing on the utility of various components. Their textbooks resembled cookbooks. In them were cataloged the recipes for hundreds of processes, separately and repetitively. The recipe for a process, such as sulphuric acid production, contained a list of equipment and procedures, with no hint that some of the procedures might also be useful in the production of, say, soda. The particularity of their descriptions hindered adaptation to other processes, so that development of new cases depended mostly on time-consuming trial and error. Then came chemical engineers to "out-science" the scientists.

Apples fall from trees and the moon rises and sets. Iron rusts and coal burns. Hundreds and thousands of such phenomena painstakingly recorded and classified would be a good start in systematic observation, but this falls short of science. Newton discovered that the same physical law governs phenomena as diverse as falling apples and planetary motion. Lavoisier discovered that the same process of oxidation occurs in phenomena as diverse as combustion and respiration. More important, these scientists elucidated the underlying natural laws and processes, and represented them precisely—sometimes mathematically —so that others could predict phenomena never seen before. To abstract from minor details, discern important patterns that sweep across disparate phenomena, uncover their general principles, and use them to explain and predict new phenomena are core abilities of natural science. They apply equally in engineering, with the difference that the targeted phenomena are mostly man-made.

In comparison to industrial chemistry, chemical engineering turns from holism to analysis, from individuals to types. It analyzes each process and focuses not on individual cases but on types of operations common to many processes. The emphasis on types ensures that the analysis is not arbitrary, because operations of the same type occur in diverse processes. The component operations conform to general principles, which make up the unifying foundation of chemical engineering.

Davis was the first to penetrate the veil of diversity in chemical processes. As an industrial consultant and pollution inspector, he visited a great variety of factories and perceived their common factors. His book *A Handbook of Chemical Engineering* contained chapters on such topics as applications of heat and cold, separation, evaporation, and distillation. Lewis observed: "Davis's first great contribution was his appreciation of the fact that the basic scientific principles underlying widely diverse chemical industries and their operation are fundamentally the same . . . Thirty years before the coining of the term, [his lectures] presented the essential concepts of *unit operations*."[45]

Chemical reactions are the crux of chemical processing, but they must be supplemented by many other operations, such as bringing reagents into proper contact. Engineers analyze a generic process into several major steps, including preparation of raw materials, control of reaction conditions, and extraction of final products. Each step is analyzed further to reveal a combination of several unit operations, of which there are a limited number of types. Reagents may be prepared by crushing, stirring, or other means. A chemical reaction may require application or removal of heat. It may involve mixture and transformation of gas, liquid, and solid phases, in which case its operations must ensure proper contact of various phases. It may employ catalysts to enhance reaction rates and hence require operations that deploy and regenerate catalysts effectively. In any case, the reactor's temperature, pressure, liquid level, and other parameters must be properly controlled; so must the flow of fluids through it. The reaction usually ends with a mixture of substances in the reactor. Besides the target product, there may be valuable by-products, unspent raw materials that can be recycled, and wastes that must be properly disposed of. Operations to

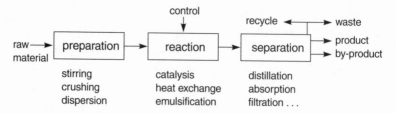

Figure 2.4   An industrial chemical process has three stages, each of which consists of a variety of unit operations.

separate them include distillation, evaporation, absorption, extraction, and filtration (Fig. 2.4).

Many unit operations, such as pressurization and distillation, are not strictly chemical. All the same, chemical engineers maintain that they deserve scientific attention because they are major hurdles in scaling reactions up to an industrial level. As a first approximation, they can be abstracted from particular reactions, developed on their own, modeled theoretically, and readied for application to a wide range of new processes.

Analysis of chemical processes into unit operations is only a first step. To pick and synthesize certain operations for a new process, engineers have to understand how and why the operations work under certain conditions. For this they need to grasp the underlying mechanisms. These have been investigated by physical chemists since the late 1880s. Walker, Lewis, and McAdams integrated physical chemistry and thermodynamics. They explained how myriad unit operations are based on a few general principles, such as fluid motion and heat transfer. They wrote in the 1923 preface of their *Principles of Chemical Engineering*: "All important unit operations have much in common, and if the underlying principles upon which the rational design and operation of basic types of engineering equipment depend are understood, their successful adaptation to manufacturing processes becomes a matter of good management rather than of good fortune." Effective prediction is the forte of science. Newtonian mechanics enables physicists to predict the return of comets. Similarly, engineers predict the behaviors of products in their design work. As McAdams explained: "The ultimate problem of engineering design is the development of methods of computation sufficiently fundamental and dependable so that from the physical and chemical properties of the chemicals involved one can predict the behaviors of new types of equipment with precision and assurance."[46] This differs from natural science only in its insistence on practicality.

From Unit Operations to Engineering Sciences

The early treatment of unit operations tended to focus on apparent physical characteristics. Its mathematics was crude. Its parameters, such as heat transfer coefficients, necessary for process design and efficiency evaluation, were empirically measured. Gradually engineers correlated empirical data mathematically and delved into unit operations to investigate their underlying mechanisms scientifically. Agitation, filtration, sedimentation, and many other operations involve fluid flow. Fluid dynamics, which studies effects of viscosity, friction, and distributions of density, momentum, and temperature, has attracted scientific attention since Newton. It is extremely complex, and its nonlinear equations resist analytic solutions except for the simplest idealized cases. Physicists turned to new loves. Engineers persevered because fluid motion is crucial in many practical applications, from aeronautics to turbine design. Fluid dynamics is especially difficult for chemical engineers, who handle not only air and water but fluids of all kinds: viscous polymers; foams and emulsions with microstructures; mixtures of gases, liquids, and solids. Even with today's powerful computers, many problems in complex flow dynamics remain to be solved. Back in the early days, progress lay not in imagining solutions but in introducing fruitful concepts that captured essential structures in empirical data, and devising practical approximations susceptible to generalization.[47]

Activity occurred in both academia and industry. DuPont established its fundamental chemical engineering research group in 1929, headed by Thomas Chilton. It included several outstanding theoretical engineers, among them Allan Colburn. Building on results of German engineers, they identified important dimensionless groups and established the analogy among fluid friction, heat transfer, and mass transfer. When the rate of one of these was known, analogy enabled one to estimate the rates of the others. Colburn never tired of seeking general concepts for engineering and explaining their significance for the design of equipment. Describing Colburn's attitude, Robert Pigford, once a

DuPont researcher, remarked: "Such clear devotion to the *use* of research results in addition to their theoretical significance is typical of early engineering research."[48]

Olaf Hougen and Kenneth Watson's *Chemical Process Principles,* published between 1943 and 1947, applied thermodynamics and chemical kinetics to chemical engineering problems. Byron Bird, Warren Stewart, and Edwin Lightfoot's *Transport Phenomena,* published in 1960, provided a comprehensive mathematical framework in distributed instead of lumped variables that covered everything from equations of changes to kinetic transport coefficients. Calling for engineers to put more emphasis on understanding basic physical principles, it had a profound influence on subsequent development of the discipline.

The scientific foundations of chemical engineering lie in several areas: thermodynamics, transport theory and fluid dynamics, reaction kinetics and catalysis. Process design is further aided by control theory and computation. Sophisticated scientific principles and powerful tools enable chemical engineers to innovate, conceptualize, design, scale up, construct, operate chemical processing facilities, and move quickly into new areas.

Scientific Engineering in Penicillin Production

The capacity for rapid adaptation was apparent in the wartime production of penicillin, which also showed the necessity of engineering science to complement natural science in producing practical results. After penicillin's discovery in 1928 by Alexander Fleming, excitement about its potential quickly subsided because no one could see a way to produce it in quantities sufficient for clinical tests. As the gathering clouds of war reminded people that wound infection was as lethal as shell explosion, a team of British scientists, led by Howard Florey, determined to make penicillin more than a laboratory curiosity. After a year of growing the mold in hundreds of pots, they extracted enough of the drug to test it on a human patient. To conserve the precious drug, they collected every drop of the patient's urine in order to extract the drug and reinject it into him. Despite the desperate recycling, they ran out of penicillin on the fifth day, just when the patient showed remarkable recovery. Because of insufficient medicine, the first patient to receive penicillin relapsed and died on March 15, 1941.[49]

An ample supply of penicillin was ready for Allied soldiers who stormed the Normandy beaches on June 6, 1944. Three months earlier, the first commercial plant for penicillin had started operation. By D-day, monthly production exceeded 100,000 doses. Wounded soldiers were injected with penicillin as a precaution against infection, and 95 percent of those treated recovered. No longer would infectious bacteria ravage battlefields and hospitals.

A potent medicine has little impact if it is more precious than fine diamonds. Increasing yield and recovery, scaling up production, and making penicillin a lifesaver was an achievement of chemical engineering. In June 1941 Florey took his research results to America. A project was promptly set up under the direction of Robert Coghill of the U.S. national laboratory at Peoria, Illinois. What proceeded was one of the greatest international collaborations in research, development, and production, involving the free flow of information among governments, industries, and universities. Scientific research improved the mold strands, feed media, and chemical processes tremendously. However, despite an elevated yield, scientists still took eighteen months to produce enough penicillin for tests on 200 patients. Then engineers swung into action, mobilizing their knowledge about chemical processing. Working closely with microbiologists, engineers at Pfizer introduced methods for deep submerged fermentation in 1,000-gallon tanks. Merck engineers invented violent circulation for oxygen transfer, a process that became the core of stirred-tank reactors suitable for growing air-breathing microbes. Racing against time, what is now called *concurrent engineering* was practiced although not named. When, nine months before D-day, penicillin moved beyond the pilot-plant stage, some firms had one team of engineers building production plants and another developing fermentation and recovery processes for the plants. The first plant to operate, Pfizer's in Brooklyn, took only five months from design to operation.

Penicillin production was a radically new process. Chemical engineers generalized their mastery of reactor principles to solve its problems, moving from tiny laboratory pans to giant industrial tanks with unprecedented speed. Furthermore, the knowledge generated in the process launched the new discipline of biochemical engineering, the first step in biological engineering, as presented in section 6.5.

Knowledge Structure and Industrial Structure

Chemicals and allied products accounted for about 2 percent of the U.S. GDP in 2000. A quarter of this value was in pharmaceuticals, a third in basic industrial chemicals, and the remainder in a wide range of products, from soaps and cosmetics to plastics, resins, paints, fertilizers, and synthetic fibers. The chemical industries constitute a major arena for chemical engineers, but in 1990 they already attracted less than one half of graduates, and that fraction continues to shrink. Besides petroleum refining and fuel production, chemical engineers also work in other industries. The food, beverage, paper, and textile industries depend on many chemical and biochemical processes; so do producers of photographic films; leather, rubber, and plastic goods; wood and lumber products; stone, clay, glass, and ceramics; metals and metallurgical products; semiconductors and microelectronics. All these, together with the chemical-production industries, constitute *chemical processing industries,* which in the United States account for about one quarter of the value of products manufactured and shipped. Despite the dazzling variety of products, their production processes share certain commonalities that make them the home base of chemical engineers.[50]

American chemical processing industries have had many success stories since the two world wars. New industries bloomed almost overnight, changing plastics and other synthetic materials from novelties into necessities of modern life. They have done well in global competition, maintaining a positive trade balance even in the 1980s, when most other industries plunged into trade deficits. They are responding and adapting to environmental issues, although they have much more to do in that area. Their rise and performance have been analyzed carefully by economists, who credit their success partly to a close collaboration between industry and academia, an integration of science, technology, and business. The chemical processing industries offer a good example of the trilateral relationship among the structure of knowledge in an engineering discipline, the structure of natural phenomena it understands and utilizes, and the structure of industrial needs it serves.

Take, for instance, the oil refining and petrochemical industries, which grew up together, with chemical engineers playing a central role in both. Crude oil pumped from the ground consists of a mixture of saturated hydrocarbons, compounds of hydrogen and carbon. Large hydrocarbon molecules must be cracked into smaller ones more suitable for motor fuels, the demand for which rose with the rapid spread of automobiles in the early twentieth century. Modern oil refining began in 1936 with the catalytic cracking process invented by French engineer Eugene Houndry. The Houndry process, which placed the catalyst in a fixed bed, required complicated equipment. Deciding to improve on it, several oil companies formed an R&D alliance in which engineers and scientists with expertise in various areas cooperated. Lewis led a team to develop *fluidization,* a powerful but complex unit operation. Taking advantage of MIT's Chemical Engineering Practice School, he brought the campus to industry and had students work under the guidance of industrial researchers to design a pilot catalytic cracker for Standard Oil of New Jersey (now ExxonMobil). It was a success for both education and industry. Fluidized catalytic cracking, in which fine catalyst particles suspend in a rising current of reacting fluid, allows continuous operation with simple equipment. It has been the dominant process in oil refining since 1945. Recalling its genesis, a Standard Oil engineer said: "One of the lessons that Doc Lewis taught us, along with teaching it to whole generations of chemical engineers, was to go for the fundamentals, to understand what you're doing, to build on a solid foundation—but not to wait until you understand everything before you're willing to make a decision."[51] That lesson is no less valuable today for all engineers.

Crude oil contains hydrocarbons, nitrogen, and sulphur, all valuable as building blocks of chemicals. Today, most plastics, resins, synthetic fibers, ammonia, methanol, and organic chemicals are manufactured with oil or natural gas as the ultimate raw material. They are called *petrochemicals,* although they can also be derived from coal or biomass, such as corn. From a few basic hydrocarbon blocks, which are byproducts of oil refining, come more than 14,000 kinds of petrochemicals. If the process to produce each had to be designed from scratch, as in industrial chemistry, the petrochemical industry would have taken a long time to build up. Fortunately, chemical engineering was ready for the task; many economic analysts find it instrumental in the rapid commercialization that made American chemical processing industries so successful.

In the synthetic rubber industry, for instance, production jumped from zero to massive in only two years, which included the development of many new technologies. Frank Howard, a patent lawyer who as head of R&D for Standard Oil of New Jersey played an active role in the development of petrochemicals, remarked: "The American synthetic rubber industry was primarily the creation of the chemical engineer. It is the chemical engineer who has given modern oil and chemical industries the equivalent of the mass production techniques of our mechanical industries." Appropriately, Howard dedicated his book about the industry: "To the chemical engineer who translates advances in chemical science into new industries."[52]

Specific chemical processes can be kept proprietary, general principles cannot. Openness is the hallmark of science, including engineering science. And it leaves its mark on the structure of the petrochemical industry through an engineering profession that emphasizes underlying scientific principles and generic operations applicable to many processes or even industries. Engineers form independent firms, where they generalize and develop the knowledge gained in petroleum refining to more complicated and varied petrochemical production. These consulting firms specialize in designing and developing manufacturing facilities for chemical processing. Some go further to perform research and develop processes universally available for licensing. Innovative and accessible in catalysts, process design, and other areas, they serve as clearinghouses of know-how and play crucial roles in diffusing technology.

Some engineering firms invest in designing complete plants with pre-engineered components, adaptable modules with certain standardized features that can be combined in various ways to suit wide ranges of requirements. Their packaged services include everything from process license, design, construction, and training operation and supervisory personnel to startup, often with guarantees on price and performance. Although low in cost, these turnkey plants use up-to-date technology. Sometimes an engineering firm develops a new design or scales up a process for a large established company, taking advantage of the operating company's experience. Then it can be hired by other companies to build a copy of that plant. In this way, cutting-edge technology is made available to the whole industry and the technological barrier to entry is greatly reduced. Small companies can enter the industry without exorbitant R&D expenditures. Oil companies can quickly integrate into petrochemicals, and chemical companies can diversify to other areas, thus realizing the economy of scope. The result is a highly competitive industry in which each important petrochemical has many producers. The impact of such specialized engineering firms was already apparent in the 1960s, when they accounted for about 30 percent of all licenses of chemical processes and engineered almost three quarters of new chemical processing plants. Similar engineering firms are spreading in other industries.

2.5 Power and Communication

Electromagnetism, one of four fundamental interactions in the universe, plays a crucial role in all physical structures. Having infinite range, it is capable of macroscopic interactions, such as carrying the sun's energy to the earth. Electromagnetic radiation can be harnessed for long-distance communication and myriad other uses. Being the predominant force in atomic structures and electronic motions, electromagnetism is also responsible for all microscopic interactions above the nuclear level, which can be harnessed for electronic devices, including electronic computers. Once its door was opened by physics, its limitless potential became the paradise of electrical engineering, which I use here as an umbrella term covering electronic and computer engineering.

Electrical engineering is like a rocket with several booster stages. The first stage was electrical power; the second telecommunications: telegraph, telephone, wireless, and radio broadcasting. Two more stages ignited after World War II, microelectronics and computers. Since the end of the twentieth century, communication, computation, and microelectronics have been rapidly converging in the information technology that some people say will transform society no less than the industrial revolutions did.

Early Electrical Industries and Electrical Engineering Education

A few electric and magnetic phenomena have been known since antiquity, but not until the nineteenth century was the inherent relationship between electricity and magnetism discovered, after efforts from many scientists. In 1831 Michael Faraday discovered the generation of electrical current by changes in magnetic intensity, which opened the way to converting mechanical energy to electrical energy and vice versa. Maxwell introduced the theory of electromagnetic waves in 1864. Twenty-four years passed before his predictions were experimentally confirmed by Heinrich Hertz, who was educated both in engineering and physics. The discoveries of Faraday and Maxwell would become the scientific bases of power and communication engineering. Physicists did not stop there and soon found themselves on the doorstep of the atomic world. J. J. Thompson identified the electron in 1897. Three years later Max Planck initiated the study of quantum phenomena. They lit the long fuses for electronics and microelectronics.[53]

Telegraphy and lighting initiated large-scale uses of electricity. In England, Charles Wheatstone built the first commercial telegraph line along the railroad between Paddington and West Drayton in 1838. It became a public sensation when the police, alerted by a telegram, waited at the train station to grab an incoming fugitive. That line was soon followed by Samuel Morse's telegraph line between Baltimore and Washington, whose operation technology quickly became dominant. In 1858, an underseas cable transmitted the first telegram across the Atlantic, and the South Foreland lighthouse threw the first beam of electric light over the sea. The telegraph and the arc light, served by dedicated electricity generators, were rather simple and did not generate large demand for electrical engineers.

By 1880, big changes were on the way. The telegraph faced competition from the telephone, which Alexander Graham Bell patented and began to produce commercially. Arc lights, which illuminated public spaces, faced Thomas Edison's incandescent vacuum light bulbs, which were more suitable for lighting homes and offices. After Edison successfully divided electrical currents, electricity was soon generated by central power stations, distributed, and sold as a commodity. William Siemens exhibited an electric railway, and electric motors of various designs joined lighting as consumers of electricity. The world was on the threshold of a four-decade spell of electrification. The electrical power and equipment industries, poised to take off, cried out for workers knowledgeable in electricity.

Universities responded, and so did students. MIT introduced the first formal course on electrical engineering in 1882. Right at its heel was Cornell, swiftly followed by other land-grant universities and more sluggishly by the Ivy League. The student body exploded. More than a quarter of the students in MIT's class of 1892 were in electrical engineering. At most universities, electrical engineering originated as a part of the physics department, and its electrical contents were mostly physics. Enthusiasm for engineering spilled over to physics. Since their inception, Cornell and MIT had required all students to take physics courses, but few students took more than the required minimum. Now electrified, they filled physics laboratories.

Sprung from the fertile ground of natural science, electrical engineering flourished under the bright sun of industry. As technology advanced by leaps and bounds, physicists who initiated electrical engineering courses increasingly found themselves unable to catch up with developments in telephony, electric lighting and motors, alternate-current power generation and transmission, and other technologies. Experienced engineers returned from industry to teach, creating distinctive electrical engineering curricula in the 1890s. One after another, electrical engineering departments acquired their autonomy. In 1884, fifteen years before American physicists formed their own professional society, the American Institute of Electrical Engineers (AIEE) was established.

Power and communications both use electrical phenomena but with different technologies. While the AIEE emphasized power, communications engineers formed their own association in 1912. Inspired by the broad sweep of the radio wave, the Institute for Radio Engineers (IRE) was transnational and open-armed from its inception, welcoming not only American engineers but also people in all nations who worked professionally with electromagnetism and electronics in other occupations. The AIEE and the IRE merged in 1963 to form the Institute of Electrical and Electronic Engineers (IEEE), the largest of the engineering or scientific professional associations.

Powering the World

The disruption of a power outage reveals how heavily modern society depends on electricity. Besides the ubiquitous electrical and electronic appliances, electrification affects economic growth by changing production processes and elevating productivity. In the age of steam engines, industrial machines were tied to a central engine and driven in groups. Electric motors freed machines from shafts and belts. The resulting unit drive enabled industrial engineers to position machines in accordance to the jobs they perform, hence to optimize their efficiency and make possible mass-production lines. Electricity overtook steam as direct industrial power in 1920. During the following twenty-five years, capital productivity in manufacturing increased by 75 percent and labor productivity more than doubled. Furthermore, the energy intensity of manufacturing dropped as electricity production became more efficient. Single-cycle energy conversion efficiency increased from 5 percent in 1900 to about 40 percent for large steam-turbine-driven generators. Total efficiency can reach 60 percent for cogeneration plants in which waste heat from electricity generation, usually in the form of steam, is harnessed for other industrial uses. Although inconspicuous to consumers, improved productivity changes living conditions profoundly by making goods and services more affordable.[54]

Electrification involves the cooperative efforts of many workers in engineering and other fields. Mechanical engineers design energy conversion engines and other machines. Civil engineers build dams to harness power from water. The technological leaders are electrical engineers. Electricity production and application are two sides of the electrification coin. Electricity supply consists of *generation*, which converts energy from fuel or water through mechanical prime movers to electricity; *transmission*, which brings large amounts of electricity under high voltage from central power plants to load-center stations in the vicinity of users; and *distribution*, which brings electricity from load-center stations under medium voltages to feeders that supply individual end users (Fig. 2.5). The use of electricity requires devices for lighting and heating; motors for industrial machines, transportation vehicles, and household appliances; current and voltage supplies for elec-

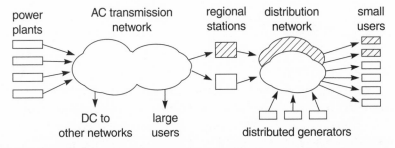

Figure 2.5   In the electricity supply system, large central power plants generate
the bulk of electricity and feed it into high-voltage alternate current (AC) transmission lines. The North American power grid is divided into four densely connected networks—East, West, Quebec, Texas—sparsely connected by high-voltage direct current (DC) lines. Regional substations step down the voltage for
distribution to smaller customers. Small generators distributed near end users
are becoming increasingly popular. Many use renewable energy sources or clean
natural gas turbines, serving specific customers and feeding the surplus into the
distribution network.

tronic devices; and meters for measurement. All these call for special
engineering design.

In the beginning, generation and application were bundled together.
Edison's system that lit the streets of New York came complete with everything from incandescent light bulbs to a power generation station.
The two sides rapidly split into the electric *utility* industry, with companies such as Consolidated Edison, and the electrical *equipment* industry, with companies such as General Electric. Commercial separation,
however, does not imply technological separation. Appliances must be
designed to fit the characteristics of available electricity, as travelers
abroad who have been frustrated by transformers and adapters know
well. Conversely, electricity suppliers must ensure that their power is
suitable to most if not all popular applications; if it is not, they will face
a truncated market and cannot profit from economies of scale in power
production. Take, for instance, the 60-hertz (Hz) power frequency now
standard in America, arrived at after engineers tried a whole series of
values ranging from 25 to 133 Hz. The final choice was somewhat arbitrary; 50 Hz works fine in Europe. But it was not too arbitrary. The
narrow range was decided by a compromise between the technical re-

quirements of electric lights and inductance motors. How the numerous players in the early days haggled in efforts to reach agreements and standards without much government regulation testifies to the force of technical rationality.

The decade-long debate on the relative merits of direct current (DC) and alternate current (AC)—so called the battle of the currents—illustrates the technical relationships between electric power and appliances, electricity generation and distribution. AC technology was more complicated. It would not have won had not John Hopkinson figured out a way to run AC generators in parallel, Nikola Tesla invented a three-phase AC motor, and Charles Steinmetz solved numerous technical difficulties. An important reason why engineers persisted in working on it, despite the early success of DC and its advocacy by Edison, was the problems in transmission. DC could not be transmitted efficiently over long distances, and this forced the location of power plants near end users, which were mostly in densely populated cities. In contrast, simple means were available for high-voltage and low-loss AC transmission, making possible the optimal location of generating plants near their energy sources. The triumph of AC highlights the importance of electricity delivery.

In many places, the user's electricity bill breaks down the costs of the service. Look at yours. Chances are you are paying less for electricity generation than for delivery.

Designing and Managing the Power Grid

Electricity supply has several peculiarities that require special engineering. It must be extremely reliable, because it is basic to most economic and social activities. Electricity cannot be stored in any significant amount and must be generated in times of use. Demand for power varies greatly daily and seasonally. To supply reliable power for fluctuating demand, the electric utility industry traditionally keeps a large margin of reserve generation capacity. To enhance economy and efficiency, utility companies interconnect their transmission networks so that those hit with excessive demand can purchase power from those with spare capacity. Large-scale networking took off in the 1920s, when engineer William Murray envisioned what he called "superpower," in which the entire nation would be densely interconnected by an electricity grid. His vision was fulfilled. Now a grid consisting of more than 670,000 miles of high voltages lines covers North America.[55]

A power grid attached to thousands of generators and millions of consumers is a gigantic network of daunting complexity. Moreover, most of the network cannot be operated with switches. Electromechanical switches are too slow for real-time flow control. Ordinary electronic switches would be instantly fried. Electronic devices capable of withstanding high voltages, expensive and rare even today, did not exist. Once electricity is fed into the grid, it spreads according to the laws of physics and the characteristics of the whole network. This phenomenon, known as *loop flow,* can send electricity on a great detour. Power from Niagara Falls destined for New York City sometimes flows through transmission lines as far away as Ohio. Grid operators have no direct control over the routes of flow. All they can do is coordinate loads and the output of the generators involved. For this, the power network is coupled to an equally complicated data network that collects and processes information critical to coordination, dispatch, and reliable grid operation.

The great difficulty in designing and controlling the power grid was revealed by the amount of engineering required. Since 1920 electrical engineers have poured tremendous efforts into modeling power grids, varying loads, analyzing flows, investigating stability, and devising control algorithms. Network analyzers were the largest commercial users of the digital computer in its early days. Their efforts contributed to the development of control and systems theories that have had wide implications.

The dense interconnection of the power grid facilitates resource sharing, but at the price of network instability and collapse. When something malfunctions, for instance when some routes are blocked by a bad transformer, loop flow automatically forces electricity into alternative lines, at the risk of overloading them and causing them to trip. Thus a minor local accident can cause other parts of the grid to break down, and this in turn spreads the failure, blacking out large areas. Such a cascade failure caused the great Northwest blackout of August 10, 1996, which had an estimated cost in excess of $1 billion. It is a heavy burden on engineers to ensure reliability, which includes the ability to supply power on demand and the ability to withstand unexpected disturbances, such as lightning striking out a transformer.

Reliability is becoming more important because sensitive information-processing equipment demands that power be steadier than ever before. However, reliability is being threatened by the recent deregulation-induced restructuring of the utility industry. Grid traffic is made more volatile by profit-hungry generators, and the behavior of each influences the performance of the whole grid. In addition, the grid is aging; many of its components have served longer than their designed lifetime. Maintenance expenditures fall precipitously because of cost-cutting impelled by market competition. A torrent of traffic is unleashed by energy trading and open access to the grid, and the complexity of grid dynamics rises steeply when traffic volume approaches the network capacity. The addition of new grid capacity almost grinds to a halt, partly because of the exorbitant costs of environmental litigation.

Power electronic switches can change the flow on a particular line, respond to disturbances in real time, and provide unprecedented control over the power grid. After almost twenty years of R&D, they are entering service. Their high price, however, may deter wide deployment in the deregulated industry now scrambling for short-term profits. To coax the aging power grid originally designed for a placid regulated industry to operate more reliably in today's tumultuous competitive market will be a great challenge to power engineers.

Perhaps history offers a lesson. Alexander Lurkis, the engineer who conducted New York City's official inquiry into its first major blackout, observed that in 1957 the presidency of Consolidated Edison passed from engineers to an accountant, who subordinated service to profit and launched a massive marketing campaign: "The program was successful as electrical demands soared but the engineering planning and decision making failed during the period from 1959 to 1981: as the disastrous 1965 blackout hit the entire Northeast; as power generator shortages forced the lowering of voltage with resulting brownouts; as power failed in large areas throughout the metropolitan area with long periods before restoration of power; as the 1977 blackout shut down the entire Con Edison system; as lower Manhattan was blacked out in 1981."[56]

Connecting the World

"Beacon to beacon rushed it to me, from Troy."[57] Within hours of the fall of Troy, the adulterous queen at Mycenae 250 miles away got the news—and the time to prepare a death trap for her victorious returning king. A similar fire-and-smoke system in another legend played a curious role in the demise of China's West-Zhou dynasty in the eighth century B.C.E. To amuse his consort, the last West-Zhou king lighted false signals for nomadic invasion and made fun of the harried lords rushing to his rescue. Guess what happened when a real invasion came. Such communication systems, prohibitively expensive, were affordable only for governments and aristocrats, but common people were no less desperate for information about loved ones far away. Images of tides and migratory birds in ancient Chinese poetry express the yearning for message carriers; so does the symbol of the letter in a floating bottle. To meet this universal need, as soon as electricity was harnessed it was applied to telecommunications.[58]

The first transatlantic telegraph message, a ninety-eight-word letter of congratulation from Queen Victoria to President James Buchanan, took sixteen and a half hours to transmit. And the undersea cable, opened in 1858, failed after two months and 732 messages, partly because of inexperience in operation. The first telephone message—"Mr. Watson, come here; I need you"—sent by Bell in 1876, arrived intact in the adjacent room, but Bell had to wait thirty-nine years to repeat the message at the opening ceremony of the first transcontinental telephone line. And he did not live to repeat it again at the dedication of the first transatlantic telephone cable in 1953. "S" was all that got through in the first transatlantic wireless message in 1901; an untimely gust blew down the receiving antenna, a wire held up by a kite. The first connection by packet switching, which would underlie the Internet, was twice as lucky in 1969; two letters, "LO," were received before a computer crashed. From telegraph to telephone to wireless to the Internet, telecommunications have advanced tremendously, but every step has been fraught with problems whose solutions testify to the ingenuity and perseverance of engineers.

The general task of communication engineering is to provide facilities for sending a message from a source to a user via a transmission system, using electric pulses or electromagnetic waves. Guglielmo Marconi, a self-educated youth who later received the Nobel Prize in physics, patented a scheme for wireless telegraphy in 1896 and brought it to wide use through the engineering synthesis of science, design, and business leadership. Ten years later Reginald Fessenden generated the continuous electromagnetic wave for wireless telephony. Wireless communication required radically new devices for generation, detection, modulation, amplification, tuning, and filtering. Some problems it shared with long-distance wired telephony. The technological challenges excited many people. Among their inventions was the three-electrode vacuum tube that generates, modulates, and amplifies waves. The vacuum tube amplifier solved the problem of repeating telephone messages over long distances, thus making possible the first transcontinental line in 1915. The age of electronics was born in the advancement of telecommunications.

Telecommunications may be roughly divided into two types: broadcast, such as television, and point-to-point, such as telephony. Bell originally envisioned the telephone as a broadcasting device, as he demonstrated in remotely delivered lectures. Radio, broadly defined by a 1906 international conference to mean communication through space without wires, was first used for point-to-point communication, especially between maritime parties. When station KDKA of Pittsburgh initiated commercial radio broadcasting in 1920, it was called wireless telephone. Its instant popularity opened the age of mass communication, which received another big boost when the British Broadcasting Company started television service in 1936. For decades, broadcasting dominated wireless communications, pushing radio telephony into niche markets such as overseas or police calls. Not until the 1980s, when satellites and cellular networks facilitated mobile telephony, did wireless rebound as a popular form of point-to-point communication.

When radio celebrated its fortieth birthday on the eve of World War II, communication engineering had achieved much in connecting the world. Telephone wires had worked their way into 40 percent of American homes, radio brought news over the air to almost twice as many, and market penetration for both was rising steadily. American Telephone & Telegraph monopolized the U.S. telephone market under government regulation. With its Bell Telephone Laboratories, it was a world leader in technology. The telecommunications industry seemed to be settling into a respectable middle age.

Few suspected that the greatest advancements were yet to come. Wartime radar research, led by the Radiation Laboratory located at MIT, developed microwave technology. That technology later set rolling the ball that broke up AT&T in 1982. The laboratory became the Research Laboratory of Electronics, MIT's first interdepartmental laboratory that brought together several disciplines, especially physics and electrical engineering. From these and other interdisciplinary collaborations emerged two related fields that became the bedrock of information technology: microelectronics and electronic computers. Electrical engineers worked closely with physicists and material scientists in microelectronics, and with mathematicians and software engineers in computers. Nevertheless, they retained central roles in both areas, which they also connected to telecommunications. After World War II, communications, together with other branches of engineering, stepped into a new historical phase.

3

ENGINEERING FOR INFORMATION

3.1 From Microelectronics to Nanotechnology

Information technology has contributed greatly to the democratization of information, which has become a commodity for mass production and instant distribution, easily accessible to most people. The major industries related to information *technology*, which does not include information *content*, are leaders in the U.S. economy. They fall into three large categories: semiconductors and other electronic components; computer equipment, software, and systems design; telecommunication equipment and service (see Table 3.1). Microelectronics, computers, and telecommunications are among the most exciting engineering achievements of the post–World War II era. Equally important is their convergence and integration in the development of the Internet and other forms of networking.

The Transistor and Microscopic Physics

"The so-called information or computer superhighway is paved with chips of crystalline silicon," observed solid-state physicist Frederick Seitz.[1] The exponential growth of computation power described by Moore's law is an achievement not of computer science but of semiconductor technology. The transistor initiated a device revolution that continues to produce ever smaller dimensions, faster speeds, higher reli-

Table 3.1   U.S. information technology (IT) industries: gross domestic income in 2000 and employment in 1998.

IT industry[a]	Income (bill. $)	Employment (000)
Hardware[b]	**105**	**836**
Semiconductors	65	284
Electronic components	21	376
Instruments	19	176
Computation	**281**	**1,979**
Computers	46	380
Maintenance and service	41	447
Packaged software	46	252
Systems design	95	548
Data processing	53	352
Telecommunication	**251**	**1,395**
Communication equipment	52	353
Communication service	199	1,042

Source: Census Bureau, *Statistical Abstract of the United States 2001*, tables 1122, 1123.
a. Excludes sales and content providers.
b. Excludes computers.

ability, and larger scales of circuit integration. Without this amazing miniaturization, we would not laugh so loudly at the 1943 assessment of IBM chairman Thomas Watson, that the world had a market for maybe five computers. How many organizations could afford a monstrous vacuum tube–based computer that would fill a building and gobble up enough electricity to light a city?[2]

The microelectronics revolution, which includes the establishment of a huge industry and a large branch of physics, is an epic in the coupled progress of science and technology. It has three overlapping phases. The first phase culminated in the invention of the transistor in 1948 and the explanation of its underlying mechanisms shortly after. This initiated a whole class of solid-state devices based on microscopic manipulation of electrons. Second was the development phase of the 1950s, and third, the technological evolution phase starting in the 1960s. Moore's law, which epitomizes the revolutionary *effects* of microelectronic devices, actually describes the technological *evolution* in the third phase. It was made possible by the accomplishments of the first two phases, among which were a deep understanding of the mechanisms underlying solid-state devices and the processes for their fabrication.

The material base of the revolution is semiconductors, of which silicon is the most famous example. With a gross income that exceeded $65 billion in the year 2000 and is growing quickly, the semiconductor industry is now among the world's largest. The bulk of its products are computer and other signal-processing chips. Besides microelectronic devices that manipulate electrons, semiconductors can be made into photonic devices that manipulate light and optoelectronic devices that manipulate interactions between electrons and photons. Although less prevalent than microelectronics, these light-manipulating devices are playing an increasing role on the information superhighway.

Semiconductors have lower electrical conductivities than metals. More important, their conductivities can be manipulated in many ways. By controlled introduction of impurities (atoms of other species), silicon can be made into an *n-type* semiconductor with negatively charged carriers or a *p-type* semiconductor with positively charged carriers of electric currents.[3] When a piece of n-type and a piece of p-type semiconductor come into contact, they form a *pn junction* that acts as a diode rectifier, passing current in one voltage polarization but not in the opposite polarization. A *pnp bipolar junction transistor* consists of a thin slice of n-type semiconductor sandwiched between two p-type layers. It has two pn junctions, configured in such a way that a small variation in bias voltage can induce a large variation in current, achieving power amplification.

Some behaviors of semiconductors were observed and put to use before they were named and their underlying physics understood. Rectifiers made of semiconductor crystals worked in wireless communications until around World War I, when they were mostly replaced by vacuum tubes. They returned in World War II, when silicon and germanium detectors far outperformed the tubes for microwave frequencies. Their revival was accompanied by an elevated status, for by then knowledge of their physics had increased tremendously.

Quantum mechanics, founded in 1925 to 1926, flung open the scientific door to the microscopic realm. It was quickly applied to all conceivable fields, including metals and semiconductors. By the end of the

1930s, a fertile bed of scientific knowledge was ready for the seeds of technology to bring forth the device revolution.

The vision was clear to Mervin Kelly at AT&T's Bell Laboratories. Having headed the vacuum-tube department and being well aware of the tube's undesirably high power dissipation and short lifetime, he was convinced that the future lay with solid-state electronics. He promoted solid-state physics after becoming Bell Labs' director of research in 1936. Soon Russell Ohl discovered the effects in silicon of what he called the pn junction. Further incited by the war effort, Kelly reorganized Bell Labs' physical research department at the first sign of peace. It would contain a key research group on solid-state physics, with a subgroup dedicated to semiconductors, both to be headed by William Shockley.

With the mission to find solid-state amplifiers and other devices, the semiconductors group made a strategic decision based on Shockley's dictum of "respect for the scientific aspects of practical problems as an important creative principle in industrial research."[4] Instead of tinkering with the complicated materials in current use to make quick and minor improvements, they decided to explore thoroughly the physics of silicon and germanium, which are rather simple, in the faith that scientific understanding would lead to major breakthroughs. Their approach paid off. On Christmas Eve 1947, fifty years after the discovery of the electron, John Bardeen and Walter Brattain made the first metal point contact transistor, which amplified power by a factor of 330. Scornful of being second, Shockley produced a theoretical analysis of pn junctions that predicted carrier injection over a potential barrier, leading to a bipolar junction transistor, which was experimentally realized by John Shive. The transistor was announced to the world in June 1948. Its underlying physics was explained in Shockley's *Electrons and Holes in Semiconductors,* published two years later.

Developing an Invention into Useful Products

The transistor was invented by physicists. Turning the invention into a revolution with wide practical applications required the collaboration of many disciplines. Chief among them were physics, material science and engineering, and at the center, crystallizing diverse knowledge into practical devices, electrical engineering. The traditional bastion of electronics, electrical engineering expanded to cover microelectronics and quantum electronics. In 1952 the Institute of Radio Engineers organized the Solid State Research Conferences, which served into the 1960s as the major forum for announcing important new results and disseminating technology.[5]

The first transistors were plagued by irregular performances, low speeds, short lifetimes, and less than 20 percent manufacturing yields. During the 1950s they were turned into a reliable and inexpensive commodity with nose-diving prices and skyrocketing speed and performance. Complementary devices for other functions were developed, and many devices were integrated into monolithic circuits that could be manufactured more efficiently.

The microelectronics revolution was a pincer of products and processes. Product development culminated in *integrated circuits* and process development in the *planar processes* for semiconductor fabrication. This two-pronged progress was accompanied by a frontal attack on the basic science and supported by strong industrial buildup. Solid-state physics grew until it became the largest branch of physics, providing a good example of technology driving science.

Befitting its roots in science, from its inception the semiconductor industry was exemplary for its openness in sharing knowledge and the mobility of its personnel. IRE's Solid State Conference emphasized disclosure and refused to invite back observers who withheld their own discoveries. AT&T was equally liberal, partly because of the regulation on its telephone monopoly and partly because it figured it could benefit from a large and dynamic industry, so long as it was in the lead. It ran seminars to transfer knowledge and was ready to license the transistor to anyone for a mere $25,000. While its Bell Labs forged ahead on semiconductor research and development, it hired carefully chosen young doctorates who, after a few years of learning state-of-the-art technology, fanned out to other companies, taking their acquired knowledge with them.

In 1955 Shockley left Bell to start his own transistor manufacturing company in Palo Alto, California. There he found an ally in Stanford University's Frederick Terman, who was working feverishly to advance engineering science. Shockley traversed the country to recruit scientists and engineers. Eight of his recruits, seven of whom had had no semiconductor experience before joining Shockley Transistor, left in 1957 to found Fairchild Semiconductor. Two of them, Robert Noyce and Gordon Moore, left Fairchild in 1968 to start Intel. Such moves were repeated. In twenty-five years, former Fairchild employees spawned more than a hundred start-ups in semiconductor devices and fabrication equipment. Thus was Silicon Valley born.

The semiconductor industry maintains close ties with university researchers. Moore explained how Intel limits internal R&D to what is needed to solve immediate problems and looks to universities for much of relevant basic research. For instance, university research produced scientific knowledge of plasma etching, and industry deployed it to replace wet etching and improve integrated circuit fabrication.[6] The semiconductor industry in the United States benefits greatly from its tradition of close collaboration among workers in all relevant fields, from physical research through chip design to equipment manufacturing. It has managed to pull together, pool R&D resources, and regain supremacy after temporarily losing ground to Japanese competition in the 1980s.[7]

As the semiconductor industry developed, product and process innovations raced forward, pulling and pushing each other. On the product side, many new devices and new device configurations were introduced. Among these was the metal-oxide-semiconductor field effect transistor (MOSFET), first fabricated by Bell's M. M. Atalla and D. Kahng in 1959 and now the most common element in integrated circuits. The integrated circuit (IC) itself is probably the most consequential invention since the transistor.

When the transistor first appeared, it was used mainly as a replacement for the vacuum tube. However, engineers soon realized that its potential far surpassed that of the tube. Being an inherently small device, it can operate at high speed and low power, both desirable. It can provide miniature implementation of large and complex systems, such as microprocessors for electronic computers. Exploitation of this potential requires miniaturization of not only individual transistors but also whole systems of many transistors and complementary circuit elements. Such was the thinking of Jack Kilby, who in 1958 invented the IC by producing resistors, diodes, capacitors, and transistors on a single piece of silicon. Kilby connected devices on his IC with soldering wires. Noyce had a better idea for connections, using the planar configuration newly invented by Jean Hoerni. Noting that all contacts in planar transistors sit on a single surface, he suggested that wires between devices could be replaced by aluminum deposition on the surface, thus making a truly monolithic IC.

Kilby and Noyce worked hard to convince people that putting many devices on a single chip would not hurt manufacturing yield. Despite initial doubts, IC chips began to roll off production lines less than three years after their first conception. That speed was a triumph of process engineering. As we saw in the context of chemical engineering, mastery of the scientific principles that underlie manufacturing processes enables process engineers to adapt them to novel products. The integrated circuit provides another example. Kilby explained: "It [the IC] could draw on the mainstream efforts of the semiconductor industry. It was not necessary to develop crystal growing or diffusion processes to build the first circuits, and new techniques such as epitaxy would be readily adapted to integrated circuit fabrication."[8] Commercial production of ICs could proceed so quickly only because it capitalized on the huge store of knowledge on semiconductor processing accumulated in ten years of struggle to manufacture transistors.

Production of uniform and reliable transistors requires ways to grow pure single-crystal silicon, introduce exact amounts of impurities, produce precise device configurations, and protect finished devices from deterioration in use. Making transistors only a few microns wide requires introducing thin layers of impurities, fabricating tiny metal contacts, and controlling all physical dimensions precisely. Toward these goals, many processes were developed and explained: diffusion, ion implantation, thin film deposition, alloying, epitaxial growth, oxide masking, photolithography, etching, oxide surface passivation. These are *unit processes* for semiconductors, to take the term from chemical engineering. Many unit processes involve complicated physical and chemical properties of solids and their interactions with various agents. Understanding and controlling them consumed much scientific and engineering research. Once mastered, they became general-purpose processes ready for use in fabricating devices of all designs and configurations, including integrated circuits.

The planar configuration is crucial to both the IC design and its pro-

duction process. In older configurations, a transistor's emitter and col-
lector contacts sit on the opposite sides of a silicon wafer. To fabricate
it, the wafer has to be flipped over and the patterns on its two sides
painstakingly aligned. This difficult procedure is eliminated by the
planar configuration, which is susceptible to the *planar process* of production. In a planar process, many unit processes are carried out successively on a single surface without disrupting alignment. It makes
possible high-throughput production of devices with complicated configurations. Designers and manufacturers can mix and match unit processes to suit their particular products. Furthermore, the fabrication
processes are independent of the patterns on the chip, which embody
the chip's logic and hence define its peculiar characteristics as a product. Pattern independence implies that, apart from broad constraints
of interface communication, designs of products and production processes can proceed on their own. It allows an efficient division of labor
between computer and semiconductor engineers. Among the consequences was the rapid commercialization of ICs, which simply means
new patterns, so far as fabrication processes are concerned.

After a decade of development, the product and process prongs of
the pincer rendezvoused to produce the device revolution. In 1961,
when Fairchild shipped mass-produced highly reliable 10-MHz transistors for the first scientific supercomputer, the semiconductor industry
passed the billion dollar mark in worldwide sales. It had matured to the
phase of technological evolution.

Achievements of Technological Evolution

Moore observed in 1965 that all basic physics problems regarding the
transistor and the IC had been solved: "It is not even necessary to
do any fundamental research or to replace present processes. Only the
engineering effort is required."[9] Engineering innovations focused on
reducing cost and expanding quantity in the 1960s, raising quality and
manufacturing yield to almost 100 percent in the '70s and '80s, and
shortening time to market in the '90s. They produced numerous new
technologies, such as the 1967 one-transistor cell for dynamic random
access memory (DRAM) and the 1973 scaling law for IC design, both
by Robert Dennard. These overall achievements are made famous by
Moore's law.

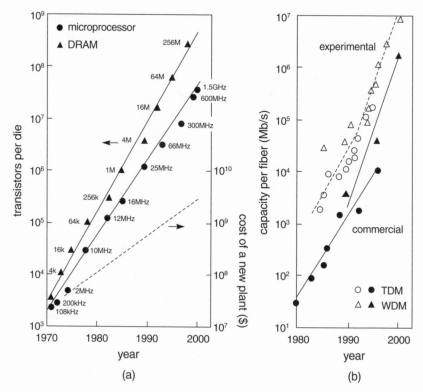

Figure 3.1   (a) The triangles and circles represent, respectively, the number of transistors per chip for DRAM, marked by memory capacity, and Intel microprocessors, marked by maximum clock frequency, both attained in the year of introduction. Their growth is described by Moore's law. The dotted line gives the cost of a new manufacturing plant, which is described by Moore's second law. (b) The information capacity of optical communication fibers has grown just as fast. Wave division multiplexing (WDM) aggregates hundreds of channels into a single fiber, and each channel uses time division multiplexing (TDM) to aggregate hundreds of thousands of voice signals.   *Source:* (a) G. E. Moore, *Optical/Laser Microlithography VIII: Proceedings of SPIE* 2440: 2 (1995) and intel.com. (b) A. E. Willner, *IEEE Spectrum* 34 (4): 32 (1997) and nortel.com.

Extrapolating from existing data, Moore predicted in 1965 that the complexity of integrated circuits, manifested in the number of minimum-cost transistors per chip, would double about every eighteen months. This has come to be called Moore's law, which is satisfied by products industrywide, not just Intel microprocessors as illustrated in Fig. 3.1a.[10]

Moore's law does not express a revolutionary breakthrough; that had already been accomplished, as Moore remarked. Rather, it expresses the incremental improvement that is normal in engineering. Engineers Murphy, Haggan, and Troutman remarked that Marx's maxim, " 'Evolution carried far enough becomes revolution,' fits IC history like a glove."[11] The effects it describes are revolutionary, but Moore's law itself expresses technological evolution. Its higher rate of increment, sustained for a longer time, distinguishes it from other engineering advancements, quantitatively but not qualitatively.

Two factors contribute to Moore's law: increasing the size of silicon wafers and decreasing the feature size of the devices. No sooner had the process of miniaturization begun, than limits were predicted by doomsayers. Successive hurdles have so far been quashed by the relentless evolution of lithographic technology, which engraves finer and finer patterns on bulk materials with light of shorter and shorter wavelengths. Nevertheless, many people argue for the necessity of revolutionary technologies, which are more glamorous than evolutionary ones, if nothing else. Paolo Gargini, chairman of the International Technology Roadmap for Semiconductors, observed that such claims caused much wasted time in the 1990s. "Fortunately," he wrote, "the manufacturing engineers, both at the suppliers and at the IC companies, pushed the *evolutionary* approach beyond any expectations and *saved* the day for *the whole* Semiconductor Industry."[12]

The first commercial ICs had a minimum feature size of 25 μm (micrometer, or 10^{-6} meter), comparable to a large bacterium but smaller than the diameter of a human hair, which ranges from 50 to 100 μm. The minimum device feature in the Pentium IV processor, shipped in 2000, is 180 nm (nanometer, or 10^{-9} meter). The new extreme ultraviolet lithography that is coming on line has produced working transistors with a feature size as small as 15 nm, comparable to a medium-size molecule. Industry predicts it can follow Moore's law for at least ten more years, although the technical roadblocks that must be overcome will increase.[13]

Nanotechnology: Revolution for the New Century
Back in 1959 when Noyce made the first monolithic IC, physicist Richard Feynman argued that nothing in physical laws precludes devices

that are only ten atoms across, roughly the size of a billionth of a meter, or a nanometer: "There's plenty of room at the bottom," he said.[14] Lithography has carried the downward march a long way, but everyone, especially the electrical engineer, knows that it will eventually be stopped by physics and economics. Physically, quantum effects such as tunneling through the insulating layer in transistors constitute an ultimate limit. Economically, the depressing news is what is called Moore's second law: The cost of a new IC fabrication plant increases exponentially, partly because of the stringent control required for high-resolution lithography. Industrial structure suffers when none but a handful of large semiconductor foundries can afford multibillion-dollar plants.[15]

Combined physical and economic constraints indicate that the evolution of conventional IC technology cannot be pushed too much farther. Some revolutionary technology must take over to continue miniaturization. The coming revolution has acquired the name *nanotechnology*. It targets the region from 1 nm to about 100 nm, a range that covers the size of most chemical and biological molecules. Structures of these dimensions exhibit many novel properties because of small size confinement, high surface-to-volume ratio, and quantum effects. They will be useful for better solar cells, lighter and stronger materials, and information storage with unprecedented density. Materials with nanometer pores can serve as catalysts in oil refining and filters that remove environmental contaminants. Devices with nanometer features designed to manipulate individual DNA molecules can improve the efficiency of genetic engineering. Nanometer structures have great potential for applications ranging from electronics to the environment, health care to space exploration. Decades of pioneering research and the development of advanced instruments have established many bridgeheads. The time is ripe for a concerted R&D effort to understand their properties, find ways to process them, and use them to design and manufacture high-performance devices with features on the nanometer scale. The U.S. government unleashed the National Nanotechnology Initiative in 2001. Other nations have similar programs.

Nanotechnology attacks the nanometer realm from two sides. From the top, evolution continues to sculpt structures with smaller and smaller sizes from bulk materials. Lithography and other solid-state technologies are applied to many other areas besides ICs, such as microelectromechanical systems (MEMS), which integrate electronics with movable mechanical parts. Because the mechanical parts can serve as physical sensors and actuators, MEMS find wide application as control devices. An old example is the accelerometer that controls the inflation of automobile airbags. A newer example is the mirror array that switches light in optical communication systems. Other exciting applications occur in biology and medicine, for instance as microcapsules for controlled drug delivery or "labs-on-a-chip" for handling DNA molecules and making diagnostic tests. Most MEMS are now on the micrometer scale, but NEMS, nanoelectromechanical systems, are appearing.[16]

The bottom-up approach, in which a device is designed and built atom by atom, molecule by molecule, is truly revolutionary. Already, instruments such as the scanning tunneling microscope can move individual atoms. Using such a microscope, engineers at IBM arranged thirty-five atoms to form their company's logo less than 5 nm in height. This method is far too clumsy to make anything commercial, however. The challenge to nanotechnology is to design the desired structures and find the process to synthesize them in mass quantity with precise control and at a competitive cost. Chemists have a leading role here, because synthesis of molecules is their forte. They have to collaborate with other experts, however, because the molecules must be further assembled into solid materials or specific configurations.

The bottom-up approach generally consists of two steps, which are sometimes performed in the same process. The first is to design and synthesize the building blocks, the constituents of which are held together by covalent or other strong bonds. The second step is to assemble the building blocks, using hydrogen or other weak bonds, to form a device. Such processes occur in nature, where strong forces bind many atoms into a small molecule, and weak interactions bind many small molecules into a supramolecule. Many biological molecules are supramolecules, and they assemble themselves. Nanotechnology aims to find similar self-assembly processes to make artificial devices.[17]

A lofty goal in molecular electronics is to create a transistor made of a single molecule and a computer processor assembled from many such molecules. It is a high hope, but inroads are being made. An example is the carbon nanotube, a remarkable molecule that is crystalline, extremely strong and stiff, an excellent conductor of heat and electricity, and can be made into a metal or a semiconductor. Since its discovery in 1991, it has attracted a frenzy of research. Transistors and small logic circuits based on nanotubes have been demonstrated. A diode has been made from a single nanotube that is metal on one end and semiconductor on the other. However, although the properties of carbon nanotubes are quite well understood, the processes for growing nanotubes with the desired properties are uncertain, not to mention the ways of assembling individual nanotubes into complicated circuits. Although a front runner in molecular electronics, the carbon nanotube has many high hurdles to jump before it can take over the baton in the race for electronic miniaturization.[18]

3.2 Computer Hardware and Software

The impact of semiconductor technology would be greatly diminished if no way existed to use the increasingly complex IC chips. But the integrated circuit has met its match in the digital computer, whose demand for IC performance is insatiable.

From abacus and algorithm, the device and the recipe for calculation, two streams converged through engineering and mathematics to form the river of computer science. The word *algorithm* derives from the name of Muhammad Al-Khwārizmī, an early-ninth-century mathematician from a Persian province in central Asia. Al-Khwārizmī wrote two books of lasting influence. The first, now lost, addressed arithmetic and was instrumental in introducing to Europe the Hindu numerals that we now use and erroneously call "Arabic." The title of the second book, which solved equations by *al-jabr* and *muqābalah* (restoration and reduction), brought the word *algebra*. Together they adumbrated three ideas central to computer science: information representation, operational procedure, and abstract symbolic manipulation.[19]

Numerals are *representations* of numbers. Many numeral systems represent the same number system, but not all are equally good for calculation. Try to do a simple multiplication with Roman numerals—for instance, CXXII times CIX—and you will understand why the Romans clung to counting boards and why good representations are so important to calculation. The superior Arabic representation, with its positional notation and its symbol for zero, facilitated the development of procedures for calculating with written numerals, such as 122×109. These new arithmetic procedures, which over time displaced the abacus and the counting board, were called algorithms. The meaning of the word gradually expanded to include other calculation routines. But that was not all. Algorithm also meant the practice of algebra, the power of which resides in the manipulation of abstract nonnumerical symbols, including symbols representing the unknown. In modern computing, algorithms extend beyond number crunching to general problem solving by operating on arbitrary symbols.

Advent of the Computer

Although the term algorithm came late, various procedures for solving specific problems had been used by people all around the world since antiquity. A famous example is Euclid's algorithm for finding the greatest common divisor of two integers. Many particular algorithms were added as mathematics advanced and problems seeking solutions increased. In the 1930s, mathematical logicians Alan Turing, Alonzo Church, and Kurt Gödel crystallized the *concept of algorithm* as a procedure consisting of a finite sequence of unambiguous instructions that transforms a given input into a specific output. They delineated what can and cannot be obtained by an algorithm—in other words, what can be effectively computed. Their work set the logical foundations for computer science.[20]

The ancient abacus and the slide rule, invented in the 1620s, are only two of many manual reckoning devices. Manual operations became more and more difficult as increasingly lengthy calculations were required to solve complex planetary trajectories and other scientific problems. Beginning in the seventeenth century, many attempts were made to build automatic machines for arithmetic operations. None was very successful. To facilitate calculation, in the eighteenth century large tables of logarithmic, trigonometric, and other mathematical functions were published. Unfortunately, they were full of errors. Enraged by the inaccuracy, Charles Babbage in 1822 persuaded the British government to finance the construction of a "difference engine" that would perform addition and subtraction with a single algorithm. During the decade-long project, he came up with the plan for an "analytic engine," a revolutionary general-purpose computer programmable by punched cards. In programming and documentation, he was assisted by Countess Ada Lovelace. Although Babbage's analytic engine was never built, some of the ideas behind it influenced Howard Aiken, whose electromechanical Mark I, the first American general-purpose computer, started operation in 1943, with Grace Hopper as the first programmer.

Technology advanced greatly in the century separating Babbage and Aiken. Electromechanical and electronic devices appeared, automatic control mechanisms were better understood, and punch-card equipment for calculators was manufactured by companies such as International Business Machines (IBM). Starting in late 1930s, Konrad Zuse in Germany and George Stibitz at Bell Labs produced computers with electromagnetic relays. Around 1941, John Atanasoff at Iowa State College envisioned an electronic computer and made a prototype called ABC. He was diverted by war duties before he could complete his work. However, he talked to Presper Eckert of the Moore School of Electrical Engineering at the University of Pennsylvania, which would become the birthplace of two revolutions that inaugurated the modern computer age: the electronic computer and the stored-program computer.

Leading a team of engineers in the Moore School, in 1943 Eckert and John Mauchly embarked on the construction of Eniac (electronic numerator, integrator, analyzer, and computer). Operational in 1946, the world's first large-scale, electronic, general-purpose digital computer contained 18,000 vacuum tubes and weighed 30 tons. With its 20 registers, each capable of holding a 10-digit decimal number, it could perform 5,000 additions or subtractions per second, about a thousand times faster than any other existing machine. It computed fast, but setting it up for a specific computation was slow. Programming was done manually by setting thousands of multiposition switches and connecting a forest of jumper cables on huge plug boards.[21]

Eckert and Mauchly were aware of their computer's weakness. In the first months of 1944, when they began to plan a follow-up machine named Edvac, they got the idea that operation instructions could be stored in the same memory devices of the computer as numbers—the concept now known as the *stored program*. Their work was classified in war time, but they discussed their plan and design in detail with John von Neumann, a mathematician who visited them in September 1944. Based on this and subsequent conversations, to which he contributed, von Neumann drafted a report on Edvac that contained a description of the logical design for the stored-program computer but no acknowledgment to the Moore School team. The draft was widely circulated in 1945, without von Neumann's permission but with him as the sole author. He never claimed explicitly that the stored program was his idea nor conceded publicly that it was not. The stored-program computer became known as the "von Neumann machine."[22]

With its gag order lifted by peace, the Moore School offered a summer course in 1946 on the construction and use of computers. Attended by twenty-eight students from twenty institutions, it stimulated a rush to construct electronic stored-program computers at many institutions. Furthermore, it imparted a broad vision for the computer's roles in society. Maurice Wilkes, who attended the summer course, recalled: "We saw computers as coming to play a central role in both science and business."[23] Hitherto, all computers had been designed for the heavy number crunching required for solving scientific problems such as calculating ballistic trajectories. Business computers for data processing were pioneered by Eckert and Mauchly. The two formed the first computer startup firm in 1946 and later sold it to Remington-Rand, where they designed Univac, the first business computer.

Besides scientific calculation and business data processing, a third avenue of application is computation embedded in physical systems. Embedded computation was pioneered by MIT's project Whirlwind, originally intended for simulating aircraft control and stabilization. Its *real-time computation* became indispensable for embedded applications such as air-defense radar-signal processing and space mission control. By 1951, the Whirlwind computer was operational and Univac began to process data for the Census Bureau. Thus began the great proliferation of computer applications.

The convergence of engineering, mathematics, and applications created a new technology. Recognizing its great potential, professionals mobilized swiftly. The American Institute of Electrical Engineers in 1946 formed what is now the IEEE Computer Society (IEEE-CS) with some 100,000 members. This was followed a year later by the more mathematically oriented Association of Computing Machinery (ACM), which has 80,000 members today. While these societies organized conferences to promote and exchange technological ideas, the computer industry mushroomed. Established firms, such as IBM and Burroughs, jumped in. IBM's dominance in the industry since the mid-1950s did not deter engineer-entrepreneurs from forming new firms such as Wang Laboratory, Control Data, and Digital Equipment Corporation (DEC). These startups and the many that followed, as much as research in universities and industrial laboratories, kept the industry innovative, competitive, vigorous, and influential.

What Is Computer Science?

While industry and professional organizations raced ahead, academia hesitated. Universities offered a miscellany of computer courses, mostly in electrical engineering and mathematics departments, and their work tended to be viewed as computer engineering. Not until 1962, when IBM's U.S.-based annual revenue from computer products exceeded $1 billion, did Stanford and Purdue take the lead in forming autonomous computer science departments. A theoretical core for the new discipline gradually formed around finite automata, formal syntax, and complexity. Computer science received further impetus from the publication in 1968 of Donald Knuth's *The Art of Computer Programming*.[24]

One reason behind the delay in forming an academic identity, observed historian Seymour Pollack, was that "the people who were claiming existence of a separate discipline called 'computer science' or 'information science' were hard pressed to identify the characteristics or cornerstone of such a science." The pressure was not eliminated by the formation of autonomous departments. "'What is computer science?' is argued endlessly among people in the field," observed R. W. Hamming in his 1968 ACM Turing Award lecture. "We call the field 'computer science' but I believe that it would be more accurately labeled 'computing engineering' were not this too likely to be misunderstood." Like Hamming, Frederick Brooks vacillated but finally called the department he founded at the University of North Carolina "computer science." Hindsight of three decades, however, changed his mind, and he wrote in 1996: "I submit that by any reasonable criterion the discipline we call 'computer science' is in fact not a science but a *synthetic,* an engineering, discipline. We are concerned with *making* things, be they computers, algorithms, or software systems." The misnomer can mislead, he warned, not the least by implying acceptance of stereotypes in which engineers have lower status than scientists. More important, it may tempt workers or misdirect students to seek novelty for novelty's sake, shun practical problems, and forget the real needs of computer users.[25]

For some time computer science was identified with programming. A 1988 joint ACM and IEEE-CS task force on "computing as a discipline" found this identification no longer valid because of the field's expansion. A satisfactory new conception, however, does not come easily. What is computer science? Are we scientists or engineers? These questions persist in the computing community three decades after Hamming's lecture. So does ambivalence. Theoretician Juris Hartmanis contended that "computer science is concentrating more on the *how* than the *what,* which is more the focal point of physical sciences. In general the *how* is associated with engineering, but computer science is not a subfield of engineering. Computer science is indeed an independent new science, but it is intertwined and permeated with engineering concerns and considerations." Agreeing with Hartmanis that attempts to separate computer science and engineering are counterproductive, Michael Loui argued this is because engineering and science are inseparable *in general:* "Engineering disciplines have a scientific basis—the engineering sciences: statics, dynamics, mechanics of solids, thermodynamics, fluid mechanics, and so on. Since World War II, all engineering disciplines have matured by incorporating their scientific foundations . . . The fundamental concepts and principles of computer science are rooted not in the physical phenomena of force, heat, and electricity, but in mathematics. Computer science is therefore a new kind of engineering."[26]

Of the top ten Ph.D.-granting computer science departments in *U.S. News and World Report*'s 2002 ranking of U.S. universities, one claims a school of its own and eight sit in the school of engineering. The remaining university has one computational department in the school of engineering and another in the school of natural science.[27] Computer science departments in many other universities and liberal arts colleges are located in schools of science and humanities. These diverse locations indicate how sprawling a field it is.

Areas of Computer Science and Engineering

Topics in computer science and engineering have been classified in many ways. Here they are roughly divided into three overlapping areas that orient, respectively, toward mathematics, engineering, and various disciplines using heavy computation: *foundations, systems,* and *applications* (Fig. 3.2).[28]

Adhering closely to its mathematical root, foundations research mainly addresses theories of computation, computability and complexity, formal models, algorithm analysis and classification, language expressiveness, and related mathematics, such as logic and combinatorics. Although some of its advanced results are remote to computing technology, it contributes to many areas, including cryptography, designs of computer languages, and practical protocols.[29]

Applications of computers are countless: information processing, storage, retrieval, and management; graphics, natural language processing, and other human-computer interfaces; artificial intelligence and robotics; automation and process control; computational science and engineering, such as computational fluid dynamics, computational medicine, computer-aided design. Computing is now an integral part of many sciences, but its main role there is instrumental. The human genome project would be impossible without powerful computers, but its focus is on the genome and not the computer. Confusion sometimes arises when *computational* sciences are counted as a part of *computer* science.[30]

Many applications are engineering intensive. Embedded computers serve as components of information systems, as in signal-processing chips in cell phones and control chips in industrial processing plants. They are found in wrist watches, video games, automobiles, and any number of other products. Although less conspicuous than the personal computer, or PC, their growth potential is far greater in the trend toward ubiquitous computing. Another class of engineering-intensive applications is large, special-purpose systems that involve heavy information processing, perhaps in conjunction with real-time input from

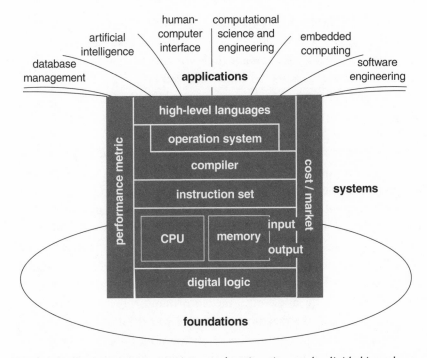

Figure 3.2 Topics in computer science and engineering can be divided into three overlapping areas, illustrated from bottom to top: foundations, systems, and applications. Computer *systems,* which include hardware and system software such as the operating system, divides into several layers of abstraction. Each layer provides service to the layer above and is implemented by layers below. It can be regarded as a "virtual machine," from whose design most complexities of other layers are hidden by carefully specified interfaces. The bottom three layers constitute the computer architecture. Of the three, the instruction set forms the interface between hardware and systems software. The layer below it is the machine organization that contains three components typical of stored-program computers: central processing unit, memory, and input/output.

sensors, for example in airline reservations and ticketing, flight scheduling, and air traffic control. The design and development of such large-scale systems pose special difficulties. To wrestle with them, a special branch called *software engineering,* the name of which originated in 1968, is becoming increasingly important.

Computing depends on computers. Providing platforms and infrastructures for diverse classes of computing applications is the job of computer systems engineering, illustrated in the center block in Fig. 3.2. Although computers are general-purpose machines, their performance can be optimized for certain classes of applications. Scientific number crunching demands high-performance floating point. Commercial data processing demands high-volume input and output. Computer engineers ascertain the functional requirements of the intended applications. Then they reckon costs, establish evaluation metrics to gauge performance, and optimize the trade-off between cost and performance.

Computer systems include both hardware and software. Systems software includes machine-user interfaces, operating systems for controlling machine operations, file managers and other utilities, and compilers for high-level languages in which application programs are written. It provides commonly useful service and support for application-specific software, such as word processors or spreadsheets. At the software-hardware interface, hiding most machine complexity is the instruction-set architecture. The instruction set presents to systems software designers all and only the information necessary to write programs that run the machine successfully. Its implementation can be considered on two or more levels of abstraction. Functionally, the machine consists of processor, memory, and input/output. These operations are implemented in arithmetic logic units, which consist of registers made up of logic gates. The instruction set and its implementation are often summarily called the *computer architecture*.[31]

The computer hardware so described is not very "hard" in the physical sense. Hardware designers do not delve into the physics of microelectronic devices. All they need to know and use are certain functional parameters, such as the clock rate of transistors, so that they can make good use of the devices. This is the *systems abstraction* typical in engineering that wrestles with enormous complexity.

Abstraction applies not only to hardware but also to software, which too is divided into a hierarchy of levels, such as systems software and applications software. None of the levels in computer systems is absolute; boundaries between them are fuzzy and fluid. This applies equally to the line between hardware and software. Many functions can be implemented in either area. Programs for specific applications can be built into chips, and, conversely, chips embedded in appliances can be designed to be reprogrammed rather than replaced. Andrew Tanenbaum writes in his textbook: "In the very first computers, the boundary between hardware and software was crystal clear. Over time, however, it has blurred considerably, primarily due to the addition, removal, and merging of levels as computers have evolved. Nowadays, it is often hard to tell them apart. In fact, a central theme of this book is: *Hardware and software are logically equivalent.*"[32]

Spread of the Computer Revolution

After this brief view of computer science and engineering, let us return to the historical evolutions of various layers of computer systems. My chronological story starts with the lower layers, which appeared first. The first machines were bare hardware, and programmers worked directly in machine languages. As machines became more complicated and engineers gained better mastery, they introduced more abstractions. Thus the layers illustrated in Fig. 3.2 appeared gradually.

If the 1946 Moore School summer course fired the starting gun for the race to build stored-program digital electronic computers, then Wilkes was the first to cross the finish line. His Edsac was operational in 1949. Although the transistor had been invented a year earlier, Edsac used vacuum tubes, because of which it was called a first-generation computer.

Second-generation computers based on transistors appeared in 1958. Systems abstraction, which allows a high-level design to be implemented in several ways, facilitated the replacement of tubes by transistors without much redesign. Architectural innovations, including multiprocessing, where several central processors shared a main memory and many peripheral processors, were introduced in 1964 with Seymour Cray's CDC 6600. This supercomputer initiated large-scale scientific computing, which has been further facilitated by successive products of Cray Research, founded in 1972.

With shipment of DEC's PDP-8 and IBM's System 360 in 1965, computers passed into their third generation, based on integrated circuits. Priced at $16,000, about one-tenth the cost of a mainframe, the PDP-8 made minicomputers competitors to mainframes. It was immensely popular in scientific and engineering communities and was often used for dedicated applications.

Most computer firms made the IC processors for their own computers. Intel in 1971 introduced the first commercial microprocessor and

marketed it as a "computer on a chip." Its success incited many followers and initiated two trends, one for specialized and the other for
general-purpose computing. Embedded computing for specific applications is achieved by specially designed chips. Microprocessors made
their way into hand calculators in 1972, and from there into all corners: home appliances, industrial equipment, information systems.
Even so, much of their potential for special-purpose computing is yet to
be tapped.

For general-purpose computers, microprocessors led to personal
computers and workstations. With their low prices, PCs appeal more
to computer novices and the consumer market. Workstations are designed for scientific, engineering, and business applications and provide
higher performance and reliability to experienced programmers, for
higher prices.

In 1975, the startup company Mits sold in kit form the first micro-
computer, the Altair. To provide a programming medium for it, another
startup was formed, called Microsoft. Two years later yet another
startup introduced the Apple II, the first fully assembled personal computer. Its popularity was catapulted by the appearance of software for
three "killer applications," spreadsheet, word processor, and database.
The fast-growing PC market enticed IBM in 1980. Bucking its tradition
of using proprietary technologies, IBM bought technologies from outside companies, including microprocessors from Intel and operating
systems from Microsoft. This "open standard" turned out to be a double-edged sword. It enabled IBM to develop its PC in record time, and
it also enabled competitors to clone IBM's products and invade the
market it had opened with the help of its reputation and promotional
campaigns. Fierce competition depressed prices and shortened product
cycles, which expanded volume and accelerated performance improvement. In the two decades after 1980, performance of microprocessor-
based machines increased 150 to 200 percent per year, while that of
minicomputers, mainframes, and supercomputers increased by about
25 percent per year.[33]

One force driving computer performance is the IC technology expressed by Moore's law. But it cannot work alone. Just as a large army
of untrained soldiers easily melts into a rout, a large number of disorga-
nized transistors generates chaos rather than performance. To organize

the exploding number of transistors on an IC and turn their increasing complexity into efficient computing is the achievement of computer architects.

The conceptual organization of the stored-program computer has three parts: processor, memory, and input/output, serving the functions of computation, storage, and communication. Besides logical and arithmetic operations, the processor also performs conditional branching, in which it controls the order of execution of instructions based on outcomes of some decision. The memory stores the program in execution together with the data it needs immediately. The input and output, controlled by the processor, fetch data from and distribute it to registers, memory cells, or other devices, which may be accessible to humans or other computers. All operations that can be performed by the three-component organization are summarized in the instruction set for systems software designers.

Under the stored-program architecture, innovative designs abound. Processing throughput is increased by parallel handling of several instructions at various stages of execution. Data access is accelerated by a sophisticated memory hierarchy that puts frequently used data in fast and readily accessible memory banks and scarcely used data in slower memory farther away. The reduced instruction-set architecture, introduced in the 1980s, further enhances speed by allowing the rapid issuing of instructions. These and other organizational designs enable a processor to find parallelism in an instruction stream, speculate on data locations and branching outcomes, choose the optimal order of execution, and thus to perform stunning computing feats at ever-decreasing costs.

Programming the Computer

Computer worship, promoted in the popular media and in some academic philosophies, makes the computer into a god in whose image people are created. A human being is nothing but a Turing machine in flesh and blood, one leading philosopher declared, because the Turing machine can in principle compute anything computable. To such empty philosophical talk, practical engineers offer the metaphor of "Turing's tar pit": even if people figured out how to program the primitive Turing machine for significant jobs, the machine would still be like a mammoth stuck in a tar pit, taking longer than the age of the universe to do what a desktop PC can accomplish in minutes. The practical problem was apparent to Hopper and Mauchly, who wrote in 1953: "The choice between two computers is, therefore, one involving the economy of operation, rather than the possibility or impossibility of such operation. For many users, one of the important economic factors is the cost associated with the programming staff needed to secure effective use of the electronic equipment."[34] They went on to argue that engineers should take account of efficient programming techniques in designing computers.[35]

Programming highlights the difference between humans and digital computers. The machine operates only on strings of zeros and ones, which to humans are torturously tedious and error prone. No sooner were the first machines up and running than engineers were trying to find programming methods more amenable to humans. Thus began the drive to *higher-level languages* and the technologies to implement them. In 1950, Wilkes invented the assembler, which translates a primitive symbolic language recognizable by humans into binary numbers for machine execution. One year later Hopper introduced the notion of a *compiler* that would enable programmers to write in high-level problem-oriented languages.

The first implemented high-level programming language, Short Code for Univac, came from Mauchly in 1952. "The most surprising thing is that an algebraic language such as this was not developed first at the mathematically oriented centers of computing activity," wrote Knuth and Pardo. Despite their surprise, the birthplace of Short Code was anything but singular. A similar idea for an algebraic compiler was developed independently by engineers working on project Whirlwind. Why it came from engineers was suggested by IBM's John Backus, who in 1954 created Fortran, the first important and lasting programming language: "Fortran did not really grow out of some brainstorm about the beauty of programming in mathematical notation; instead it began with the recognition of a basic problem of economics: programming and debugging costs already exceeded the cost of running a program, and as computers became faster and cheaper this imbalance would become more and more intolerable."[36] Practical concerns, not theoretical ones, initiated the development of high-level programming languages.

High-level languages relieve human programmers of the drudgery of supervising detailed machine operations such as assigning registers or setting aside memory for intermediate results. However, these tasks must be performed and performed effectively if the machine is to run satisfactorily. Originally the chores of human programmers working in machine code, they are delegated to the compiler, which not only translates a high-level language program into assembly language but also analyzes it and fills in the details necessary for machine execution. To develop a reliable and efficient compiler is no easy job. The one for Fortran took twenty-five man-years. Nowadays compilers can be developed much faster with the help of theories and programming tools, and they can produce machine codes as efficiently as experienced human programmers.

Among the advantages of high-level programming languages is their adaptability to the needs of various computing applications. As applications multiplied, so did languages. By 1972, more than 170 languages were in use in America alone. Object-oriented programming, which combines process and data for more powerful abstraction, led to C++ and other languages that allow software components to be reused and extended. Web-based applications led to Java and other platform-independent languages. And it goes on.

Using the Computer

Whereas compilers for high-level languages automate routine chores for *programming* computers, operating systems automate chores for *running* them. As computers became more sophisticated, human operators could no longer catch up with machine speeds. In the mid-1950s, engineers began to develop operating systems that would control the operations of a computer and manage its resources for maximum utility.[37]

Operating systems not only schedule different jobs but also ensure correct resumption of interrupted jobs and prevent faults in one job from jeopardizing other jobs or the computing system itself. Furthermore, their management overhead must not be so high as to use up all the computing power. An important achievement in the mid-1960s was OS360, the operating system for the IBM 360 family. It worked for all machines in the family, from lightweights for business data processing to heavyweights for weather forecasting and other scientific computing. Despite its sophistication, however, OS360 disappointed many programmers yearning for multiprogramming and interactive programming facilities. In multiprogramming, a computer processor interleaves several jobs so that instead of wasting processing time while waiting for the input of one job it switches to execute another job. A subclass of multiprogramming attaches a computer to many consoles, through which programmers interact with the computer, debugging their programs in real time instead of waiting for scheduled batches. This scheme, known as *time sharing,* aims to use scarce computing resources efficiently.

A prototype for a time-sharing computer serving multiple users was demonstrated at MIT in 1962. Emboldened, MIT teamed up with AT&T to build an operating system that would enable a central computer to serve a large community of users. Multics (Multiplexed Information and Computing Service) was not a practical success; it required technology far more sophisticated than its time could offer. Its main achievement was visionary. Instead of the prevailing view of *computers as products,* it introduced the idea of *computing as a service or utility* as easy to access as electricity. This vision is increasingly being fulfilled by the grid, distributed computing that is emerging now.

Although the interactive computing service envisioned by Multics failed to materialize for large communities, it did bring great convenience to the small technical groups that worked on it. When AT&T withdrew in 1969, Bell Labs engineers, among them Kenneth Thompson and Dennis Ritchie, looked for a substitute and created Unix, named as a pun on Multics. Unlike previous operating systems written in the assembly language, Unix aimed for simplicity and was mostly written in a high-level language called C, which they developed concomitantly. C enabled Unix to be transported to other machines and provided it with an easy interface for developers of application programs. Distributed among universities, with its source code, for a nominal fee, Unix quickly acquired a great following and was modified into many versions. The most important modifications came from the University of California at Berkeley in the late 1970s. Among other things, Berkeley added networking capability and Internet protocols. Instrumental for spreading Internet usage in universities, it established a strong position for Unix in Internet applications.[38]

For more than a decade, Unix was the dominant operating system for workstations, in distinction to PCs monopolized by Microsoft's MS-DOS and later Windows. Unix operating systems are more reliable, serviceable, and network accessible. However, they have a great weakness. Because Unix has many versions, it is less attractive than the unified Microsoft systems to developers of applications software. And applications are what attract most consumers.

Microsoft's source codes are a jealously guarded secret; so are most commercial Unix versions. However, the open exchange of source codes in Unix's academic days spawned the *open-source software* movement, in which source codes are made available for redistribution without charge or restriction. In this way software developers give their users more influence on product evolution and in return benefit from the open sources of other developers. To prevent the arbitrary modifications and proliferation that so weakened Unix, the General Public License was established in 1984 to stipulate legal constraints on free distribution. Under it, Linus Tovalds distributed in 1994 a high-performance and highly reliable operating system called Linux. It quickly attracted many vendors and much technical support, including such big guns as IBM. Intel slackened its bond to Microsoft and began to optimize its processors for Linux operation. Although Linux poses little threat to Windows in the consumer world, it is making inroads in the corporate world. In 2002, it grabbed about 13 percent of the $50 billion market in servers, beefed-up PCs or workstations for web servers. More important, it has great potential as an operating system for programmable embedded devices such as cell phones, video games, and personal digital assistants, which can download information and the programs to process that information through the Internet. This seems to be the wave of the future.[39]

"The new applications come from the new era we are moving into: *Post-PC* is the name David Patterson gives it," wrote John Hennessy about computer systems research in the new century. Noting that sales of embedded processors overtook PC processors in 1997 and are expected to triple those of PC processors in 2003, he wrote: "Information appliances are absolutely a big part of the future . . . It's an Internet- and Web-centric environment in which services are the killer applica- tions. In such a world, your net connection becomes more important than any one computing device."[40] In this new era, computer architects must join forces with architects of communication networks.

3.3 Wireless, Satellites, and the Internet

In his theory of communications, published in 1948, one year before the operation of the first stored-program computer, Claude Shannon had the prescience that communication and computation would build on each other. Although communication was all analog then, he fo- cused on the digital case because "this case has applications not only in communication theory, but also in the theory of computing ma- chines."[41] His focus has been amply vindicated. Communication engi- neering has revolutionized itself, assimilating and stimulating concomi- tant developments in computers and microelectronics. As the new century begins, it is at the cutting edge of high technology, busy devel- oping global information infrastructures.

Communication Systems

A generic communication system contains at least five parts: *source, transmitter, channel, receiver, destination*. A source generates a mes- sage, which if necessary is converted by a transducer into an electrical signal, the waveform of which physically embodies the message's infor- mation. The transmitter transforms the signal so that it can be sent most effectively over a certain channel, or transmission system, which is subjected to random noise that induces errors. When the signal ar- rives at the receiver, it is cleaned up and transformed into a format ap- propriate to the user, its destination. The user also specifies the desired quality of service, including a criterion of fidelity (Fig. 3.3).[42]

A communication system depends on both physical and systems technologies. Physical technologies consist of hardware: electronic de- vices for generating and modulating electrical signals, transmission me- dia with required repeaters and switches, and detectors at the receiving end. Systems technologies integrate the physical components logically or procedurally into a system serving desired functions. They divide

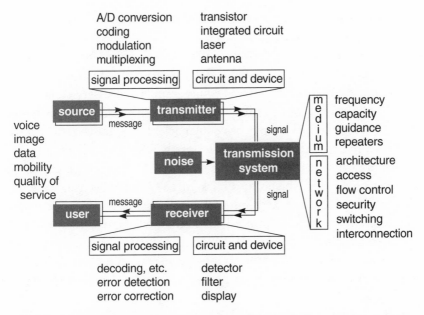

Figure 3.3 A generic communication system sends signals from sources to users via transmitters, a transmission system, and receivers. It encompasses physical and systems technologies. The former includes devices and transmission media; the latter signal processing and network architecture (A/D = analog-to-digital).

into two interrelated classes: *signal processing* and *network architecture*. Signal processing designs the forms of signals suitable for effective transmission, the procedures for transforming source signals into the desired forms, and the algorithms for error detection and message recovery at the receiving end. Network engineers take account of characteristics of signal sources to design effective connections to numerous terminal devices; procedures for granting access, switching, and controlling traffic; protocols for initiating, maintaining, and terminating data transmission. All designs aim to provide the quality of service required by intended users. Communication engineers try to maximize the volume of traffic in a system and minimize the probability of errors. They often have to trade off these aims against the complexity and cost of the required equipment.

Every part of the communication system scored breakthroughs in the postwar era. Traditional sources and users are joined by computers and mobile parties. Devices multiply with advances in microelectronics, photonics, and other computing chips. Transmission media expand to include microwave, satellites, and optical fibers. Signal processing become increasingly digitized. Architectures now include packet switching, internetworking, and cellular networks. Consequently, telecommunications has gone way beyond vanilla voice service between people at fixed stations to offer many flavors in forms and user types. Since the commercial rollout of cellular telephony in the early 1980s, the potential of wireless to provide untethered communication is increasingly being realized. Facsimile machines for transmitting images, the first patent for which dates back to 1843, have been in wide use since the mid-1980s.[43] Computer data transfer has become a major part of communications traffic. The Internet, which in its first two decades was confined to technical communities, gained its first commercial service provider in 1990. A global information highway system is emerging that offers services integrating voice, image, video, and data.

Opening New Transmission Media

All telecommunication systems use electromagnetic waves to carry signals. Unguided waves propagate in a vacuum or the atmosphere; guided waves in wires: twisted pair, coaxial cable, and optical fiber. Waves in a wide range of frequencies are used. The amount of information that can be carried by a waveform is proportional to the waveform's *bandwidth,* or range of frequencies in which almost all of its energy is concentrated. A carrier wave with higher frequency has a larger bandwidth and thus a greater capacity for information transmission. For instance, a bandwidth of around 4 kHz (kilohertz, or 10^3 Hz) is required to transmit the human voice. An open wire, like those in early telephone networks, operates at the frequency of a few kHz and has just enough capacity to carry a single voice channel. The coaxial cable, deployed since the 1940s, can operate at up to 500 MHz (megahertz, or 10^6 Hz) and carry tens of thousands of voice channels simultaneously. It is also used for cable television.

Incessant demands for larger transmission capacity push telecommunication toward carrier waves with shorter wavelength, higher frequency, and larger bandwidth. Using these waves requires increasingly sophisticated scientific knowledge. Broadcast transmission pushed up to the 100-MHz range before World War II. Beyond lay microwaves, with frequencies above a gigahertz (10^9 Hz). Microwave transmission, first demonstrated in 1931, was commercialized after the war with technologies developed for radar. Because high-frequency waves show little diffraction bending and can benefit from directional antennas, they are mainly used in line-of-sight links for point-to-point communication. The first microwave link relayed signals between television stations in Boston and New York by five hops over 220 miles. Within fifteen years, microwave links spread to cover continents. When terrestrial networks reached the shoreline, they were connected by stations in the sky.

AT&T's experimental Telstar, the first communication satellite to receive and transmit signals actively, soared in 1962 to bridge the Atlantic and carry live intercontinental telecasts. Three years later the international Intelsat 1 opened commercial satellite communication. Perched 35,800 km up in a geosynchronous earth orbit (GEO), it appeared stationary to observers on the ground and could illuminate one third of the earth's surface. Today about 180 GEO communication satellites from various nations encircle the equator. Since the 1980s, they are the mainstay for the distribution of television programs, linking program sources and network centers to local broadcast stations or cable heads equipped with large earth terminals. Direct-to-home television transmission, which requires high-power satellites to generate signals strong enough to be detectable by small home receivers, have been commercially viable since the early 1990s.[44]

For transmitting telephone conversations, satellites are less successful. GEO satellites sit so high up that even light traveling at 300,000 km per second takes a noticeable amount of time to make a round trip, causing enough signal delay to annoy human speakers. Thus satellites quickly lost their share of telephone transmissions when optical fibers became competitive in the mid-1980s.

Human eyes are sensitive to electromagnetic waves with frequencies between 390 and 770 THz (terahertz, or 10^{12} Hz), equivalent to wavelengths between 0.76 and 0.39 μm. The visible band, together with infrared wavelengths just above it and ultraviolet wavelengths just below it, is usually called *optical*. The first modern telecommunication system, the French mirror telegraph built by Claude Chappe in 1793, was optical. It relied on visibility. The biggest potential of optical communication, however, lies elsewhere. Not in the speed of light; electromagnetic waves of all frequencies travel at the same speed in free space. Its main advantage is in its enormous bandwidth that, like a fat pipe, permits high rates of information flow.

Glass fibers were used to transmit light for medical examinations in the 1950s, but their loss was so high a room-length strand was opaque. When W. A. Tyrrell argued in 1951 for the superiority of optical over microwave frequencies for telecommunications, he lamented the lack of both adequate light sources and transmitting media. Hopes for light sources were raised by the laser (*light amplification by stimulated emission of radiation*), theoretically predicted in 1958 and experimentally demonstrated by Theodore Maiman two years later. Hopes for better transmission media were raised by Charles Kao, who in 1966 demonstrated that the loss of optical fibers could be reduced to levels suitable for interoffice communications. Major breakthroughs occurred in both fronts in 1970. Semiconductor lasers achieved continuous operation at room temperature.[45] And a single-mode fiber meeting Kao's criterion was produced commercially. Fiber loss dropped steadily during the following decade to the near-theoretical limit of 0.2 decibels per kilometer (dB/km), so that 10 percent of input power remains after a signal travels through 50 km of fiber. Roads to optical fiber communications were clear.[46]

Trial systems sprang up in several cities around the world, including Chicago in 1977. The progress since then can be seen in Fig. 3.1b. With a potential bandwidth exceeding 60 THz at wavelengths between 1.2 and 1.6 μm, the silica fiber offers the fattest information pipe. It is also the thinnest in physical size, so that it easily snakes into office buildings where space is at a premium. Its immunity from interference charms the military, which worries about electronic jamming. More important for telecommunications, its low loss implies fewer amplifiers en route, its low noise implies low error rates in transmission, and its robustness implies long service life. With all these advantages, the adoption of optical fibers has been rapid. By the end of the twentieth century, optical fibers made up almost the entire backbone of the world's long-haul telecommunication network, and they are increasingly pushing their way into metropolitan and feeder loops. More than 350,000 km of "glass necklaces" grace ocean floors, replacing old copper telephone cables for undersea seismic and other scientific research. Their information transmitting capacities are so large that a glut occurs.[47]

The telecommunication network now is like a ten-lane freeway system with treacherously narrow access ramps. Trunk lines and connections to signal processing units located at street corners are optical and fast. However, the last fifty yards between a corner unit and an individual home are mostly bridged by the old twisted-pair wires designed for voice transmission. These customer connections took decades to construct and constituted about 60 percent of AT&T's total investment before its 1982 divestiture. Replacing them is exorbitantly expensive. To access many homes for broadband services encompassing telephone, television, and the Internet is difficult but potentially profitable. There is now a fierce struggle among various technologies: modifying existing twisted pairs into digital subscriber lines, providing modems for exiting television cables, new fiber cables, wireless connections to neighborhood terminals, direct satellites, and more. Communication engineers face an exciting frontier in this area alone.[48]

Signal Processing and Network Architecture

So far we have focused on physical technologies. Equally important are systems technologies that determine the required forms of signals and their paths from specific sources to specific users. An example of signal processing is the transformation between analog and digital signals. In 1937 Alec Reeves invented a scheme to digitize signals, which was commercially deployed in 1962, when mass-produced transistors provided cost-effective implementation. It measures a continuous signal thousands of times a second, quantizes the measured results into hundreds of levels, and represents each level with a binary digital code. The result is a digital signal, a sequence of on and off pulses, which can be further processed and transmitted by the same equipment that also handles computer data and digitized images. Digital signals are superior for long-distance transmission with minimum distortion; unlike amplifiers of analog signals, repeaters of digital signals do not accumulate noise. Now most trunk links between telephone central offices are digital and capable of providing integrated services.

Transmitters in a communication system handle much more signal processing than analog-digital conversion. An important process is *modulation,* which maps a signal onto a carrier wave suitable for transmission. Many modulation schemes exist besides the familiar amplitude modulation (AM) and frequency modulation (FM) for radio. At the receiver end, the signals are *demodulated* to recover the forms intelligible to users. A modem (*m*odulation-*dem*odulation) modulates digital signals from a computer for transmission over analog telephone lines.

Another important process for the transmitter is *multiplexing.* It bundles many signals for a single carrier without mixing them up, thus ensuring efficient use of available bandwidths and transmission resources. The first scheme developed was time division multiplexing (TDM), which interleaves pulses from different digital signals in consecutive time slots for transmission. This is how the single fiber in the system OC-1 with a transmission capacity of 51.8 megabits per second carries hundreds of telephone conversations simultaneously. Digitization transforms a voice signal into a string of on-or-off pulses at 64 kilobits per second. A time division multiplexor goes around 672 such strings, taking one pulse from each in turn so that all pulses are interleaved into a single stream at 43 megabits per second. Then the stream is launched into the transmission fiber. At the other end, a demultiplexor divides the stream to recover the 672 signals and send them on their separate ways. A TDM stream uses a specific carrier wavelength. Advanced optical fiber communications systems bundle many TDM streams with various wavelengths in wavelength division multiplexing (WDM). TDM and WDM are crucial to the exponentially increasing transmission rates depicted in Fig. 3.1b.

A point-to-point communication network must connect any pair of its numerous subscribers on demand, use shared resources, and ensure that the sharing does not jeopardize privacy and availability of service. *Network design* and *traffic control* are among the most difficult of telecommunication problems.

Dragging along a cord when one phones is not very appealing. Untethered and mobile telecommunications have been possible since the birth of wireless, the first application of which was to serve ships. Land mobile communications had two major obstacles. The first, lightweight transmitters and receivers with low power consumption, has been overcome by microelectronics. The second problem concerns transmission. More cables can be laid for guided transmission, but there is only one atmosphere, with a limited bandwidth, that must be shared by all unguided transmissions, and its prime estates are occupied by broadcasting services. Available bandwidth confined land mobile telephony to a handful of channels as late as the 1960s, mostly used by the police and taxi dispatchers. As bad as the services were, they had long waiting lists for subscribers. Engineers saw clearly that to satisfy the mounting demand, transmission frequencies somehow have to be reused. Unlike optical fibers that depend mainly on physical technologies, wireless solutions emphasize advanced systems technologies in signal processing and network architecture.

In 1971, Bell Labs introduced the *wireless cellular architecture,* which adds a spatial dimension to achieve multiple frequency use. It discards the old scheme in which a single high-power station covers a large area. Instead, it divides the area into small contiguous cells, each with its own low-power station. The same frequency can now be used in different cells to carry more calls, because attenuation ensures that the low-power transmitter causes minimal interference in neighboring cells. When a person talking on a cell phone moves from one territorial cell to another, the cell's station hands off the call to the station in the other cell. Cellular systems demand complex schemes to authenticate callers, locate receivers, allocate frequencies, set up and hand off calls, secure privacy, track user movements, and connect to fixed telephone systems where origins and destinations of most calls reside. Complexity has not dampened worldwide enthusiasm. A plethora of cellular systems with diverse designs appeared in the 1980s. Using analog technology reminiscent of FM radio, these first-generation cellular systems provided one-way paging and two-way voice services. Second-generation systems switched to digital technology and added data service, although the quality of service still leaves much to be desired. The number of subscribers grows 20 to 50 percent annually, more slowly in America than in Europe and Asia. Third-generation mobile radio networks, eagerly awaited at the turn of the twentieth century, strive for global consensus in multimedia wireless services with speedy Internet access.[49]

Convergence of Communication and Computation

Spectrum, IEEE's member magazine, featured a profile of "Architects of the net of nets" that began: "One is dapper, convivial, pragmatic—and from California. The other is reserved, deep in piles of information, deeper still in concepts—and from New York City. Yet through four decades the careers of Vinton G. Cerf and Robert E. Kahn intertwine 'like a historic court dance as we cycle through a *pas de deux,*' to quote Cerf. And the supreme product of their supremely productive collaboration was the Internet."[50] Pragmatic Cerf works mainly in computing, conceptual Kahn in communication. Their intellectual bent and collaboration reflect the global information infrastructure as a waltz of computation and communication on a silicon floor.[51]

A generic *internet,* as its name indicates, is a network interconnecting networks. Internets and their technologies are more general than the characteristics of the Internet that is now in global use. The Internet traces its roots to the Arpanet, which is not an internet but simply a data network designed to connect computers. Computers behave differently from people. Connecting them requires new technologies different from those for telephones, which are designed for person-to-person communication. The innovation that distinguishes the Arpanet is packet switching.

Traditional telephone networks employ *circuit switching,* in which a physical circuit is set up for the exclusive use of a call throughout its duration. The dedicated circuit suits humans; two persons talking over the phone produce a more or less continuous stream of messages, and they are annoyed if responses from the other side are delayed for a second in a telephone transmission. Unlike regular and impatient humans, computers are irregular, bursty, but tolerant of delay. A computer can be silent for hours while its human user accumulates a pile of e-mail, then burst into a flood of transmission when the user clicks "send." A dedicated line between two computers would be a great waste, because it must have high capacity to handle irregular bursts but would sit idle most of the time.

Better efficiency for computer communication can be achieved with *packet switching,* which replaces static circuits with dynamic routing. If circuit switching is like reserving a railway line for a train, then the packet switching architecture is like a highway system accommodating many cars. A message must share its way with other messages. Because long messages may hamper effective sharing, they are not treated as wholes. Packet switching divides a message into many small parts, or packets, that travel separately to the destination, where they are reassembled. At each exchange en route, a packet waits in line as it arrives and then is sent on toward its destination when its turn comes. To avoid congestion on certain routes, packets in a message may be dispatched along different routes. Because some routes are faster than others, the packets may arrive at their common destination out of order. Thus extra measures are required to ensure their proper reassembly. Packet switching enhances network utility at the price of quality of service. Packets may get lost, and their routing and reassembly may cause delays in message delivery. Delays do not bother computers but can exasperate humans users. That is why Internet telephony is so late in coming.

The idea of packet switching occurred independently to three engineers: Paul Baran in the late 1950s; Leonard Kleinrock, who published the first paper on the topic in 1961; and Donald Davies, who formally proposed switching by "packets" in 1965. Kleinrock imparted his theoretical conviction to Lawrence Roberts, whose own ideas firmed up in 1965, when he connected computers via a dial-up telephone line. His experience gave him a fair idea of the required technology when he was recruited by the Advanced Research Projects Agency (ARPA) to develop computer networking.

ARPA, which later added Defense to its name to become DARPA, kept a keen eye on the information processing situation, partly because the U.S. government paid for most of the computers. It saw two problems in the 1960s. First, the great variety of incompatible computers coming from many manufacturers held out little hope for standardization. Second, research facilities were scattered all across the country, leading to much redundancy and inefficient use of computing resources. Bob Taylor proposed to solve these problems by building a nationwide network of electronic links among the computers, so that computing resources could be shared and incompatibility mitigated by communication. His proposal was immediately funded, and Roberts, whom he picked to be program manager, arrived in 1966.

Roberts was responsible for formulating a general conception of computer networks, based on which he would solicit proposals for specific components. A major problem for networking is that the host computers to be connected use widely disparate operating systems. The communication system must make them "understand" each other. One way to achieve compatibility is to write communication software to run on specific operating systems. However, that requires many versions of the complicated software that have to be rewritten whenever communication technology changes, making network development difficult and costly. To solve this problem Roberts adopted a suggestion made by Wesley Clark. He applied the general engineering ideas of *modularization* and *abstraction* to separate the tasks of computation and communication, so that each can be modified with minimal interference to the other. In modularization, most communication tasks are taken from host computers and assigned to specially designed minicomputers called IMPs (interface message processors). The IMPs form the communication network's nodes, which hide the network's complexity from host computers. All that is required of a host connected to a node is compatibility with the IMP interface. For this it must conform to a *protocol* that defines strict rules for preparing data for packet switching and procedures for initiating, maintaining, and terminating transmission. Protocols are the computer equivalent of the procedures a person must perform to place a telephone call, and IMPs the equivalent of the mechanisms working inside the telephone. With the help of IMPs, computers are like people: we need to know very little about the telephone system to place a call, and we can switch from rotary to touchtone phones with minimal effort.

ARPA awarded the contract for designing and building IMPs to the firm Bolt Beranek and Newman (BBN). The communication theoretician of the BBN team, Robert Kahn, was also responsible for overall system design. Design of the host protocol fell to the Network Measurement Center at the University of California, Los Angeles, where Stephen Crocker and Vinton Cerf were star graduate students in computer science. Success of a protocol depended on the cooperation of participating host communities. To stimulate ideas and forge consensus, Crocker initiated *request for comments,* an informal practice that has become an influential tradition in network development. The

Arpanet was commissioned at the end of 1969, when traffic began on a leased telephone line between its first two nodes, at UCLA and the Stanford Research Institute. The network grew to fifteen nodes and twenty-three hosts in 1971 and became international two years later.

To introduce Arpanet to technical communities and stimulate development of applications for computer networks, Kahn organized the first International Conference on Computer Communications (ICCC) in 1972. With demonstrations that connected forty machines, the conference kicked off real growth in network traffic. One of its results was the formation of the International Network Working Group (INWG), with Cerf as the first president. The first of many grassroots organizations that steered the creation and evolution of Internet standards, INWG continued and strengthened the request for comments practice.

ICCC and INWG highlight a peculiarity that influences the Internet's characteristics. Unlike the telephone network, the Internet did not grow under a monopoly. It grew from a peculiar combination of top-down and bottom-up efforts. The government initiated the broad vision and general conception, but it does not control specific content and technologies. Development of Internet technologies always falls on a kind of consensus formation among technical communities with a stake in networking. This is a more democratic process, but it also generates problems, as in maintaining network security.

From Computer Network to Network Network

As networks proliferated in the 1970s, so did their variety. Various nations developed their equivalents of the Arpanet with protocols that were incompatible with each other. These wide-area networks covered large districts but had relatively low rates of data transmission. To provide a high data rate to localized communities, such as a university campus, local-area networks with disparate topologies appeared. Furthermore, new modes of telecommunication, such as wireless and satellites, demanded new networking technologies. When Kahn became an ARPA project manager in 1972, he faced a nascent field enormously rich but disorderly.

A good engineering scientist, Kahn did not jump into the specifics of hitching satellites to the Arpanet. Instead, he posed a general question about the necessary conditions for interconnecting disparate networks, whatever they are. For the general characteristics of an internet that can be readily expandable to accommodate networks based on new technologies, he introduced the idea of *open-architecture internetworking*. It extends the logic of Arpanet from a network of computers to a network of networks. In the open architecture, an internet would have no global operational control. It would not dictate the internal structures of participating networks, each of which would retain its own individuality. At the interface between two networks would be a gateway or *router*, which would ensure the compatibility of signal transmission but retain no record of the contents of the signals that pass through. To work out the details of his internetworking architecture, Kahn recruited Cerf, who had become a leading expert in communication-oriented computer operating systems. Thus began their long collaboration in the development of internetworking.

Cerf and Kahn's paper on the transport control protocol (TCP) for internetworking was published in 1974. Refined through use, the TCP split into the TCP and the internet protocol (IP) to accommodate two modes of packet switching. In 1983, TCP/IP was exclusively adopted on the Arpanet and bundled with Unix in Sun Microsystems' workstations. Despite competition, it was firmly established in the 1990s as the dominant set of protocols for global internetworking.

As is common in engineering abstraction, the logical architecture of an internet consists of many carefully interfaced modules. Together they ensure the reliable transmission of a message from one application to another, be it e-mail or a web browser, that may be located in a different network. Among the modules is the transport layer, for which TCP is the dominant protocol. To prepare for packet switching, the transport layer of a host computer divides the message handed to it by an application into small packets and adds to each packet an electronic "envelope." The envelope contains such information as the recipient application, an error detection code, and a sequence number locating the packet in the message. Then it passes the enveloped packet to another layer, which adds another envelope with further information, such as the address of the application in a distant network. Layers of envelopes are added, up to the computer's physical port designated to launch the packet into the communication network. Once on its way, the packet comes to a router that, like a border station between two na-

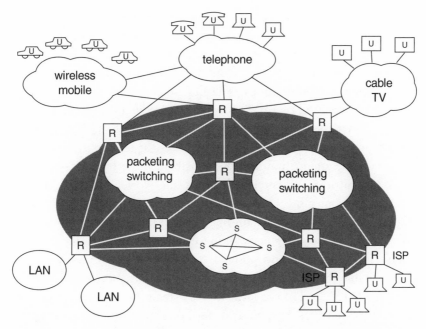

Figure 3.4 An information infrastructure consists of a core of packet-switched networks (dark area) interconnected by routers, marked *R*. Other routers serve as gateways to other telecommunication systems, such as wireless mobile networks, fixed telephone networks, and cable television networks, each providing access to many small users, marked *U*. Internet service providers (ISP) also offer access to larger users and local-area networks (LAN).

tions, connects two networks according to IP. The router notes the information on the outermost envelopes, consults the traffic situation, chooses a good route to the destination, revises the envelopes for accessing the appropriate network, and sends the packet on its way. At the destination computer, various layers strip off the envelopes, check for errors, request retransmission if necessary, assemble the packets, and deliver the message to the addressed application.

The structure of an internet is schematically illustrated in Fig. 3.4. At its core are packet-switched networks interconnected by routers. At its periphery, routers equipped with modem banks connect to various networks, for instance the circuit-switched telephone network, through which many PC users gain access.

Unlike voice communication that almost anyone can use, data communication demanded of users considerable technical ability. Users of data networks were mostly engineers and scientists. They contributed much to the development of applications, without which Arpanet and its successors would have been confined to information dissemination among research communities. The first major application to appear was e-mail. Invented by BBN's Roy Tomlinson in 1972, it made up 75 percent of Arpanet traffic in the following year. Mailing lists and news user groups followed.

Perhaps the application most responsible for the Internet's breaking out of laboratories and entering homes is the World Wide Web (www). It was created at CERN, the European Center for Nuclear Research, originally to facilitate heavy on-line data pooling by elementary particle physicists. Computer networking was pervasive at CERN in 1989 when Tim Berners-Lee proposed a distributed hypertext system that would enable the creation of links between data scattered over diverse computers. To develop the idea, he led a team of physicists and engineers to create the Hypertext Transfer Protocol (http), which supports data communication for easy browsing of information. For systematic identification of information, they introduced the Uniform Resource Locator with a standardized network address format. The web is designed to be portable and scalable, so that it can be expanded to accommodate additional information without disturbing existing contents. CERN distributed web software to elementary particle physics communities in 1991. By the end of the year, ten web servers had opened up shop worldwide. To improve browser performance, Marc Andreesen led a team at the National Center for Supercomputing to develop Mosaic, a faster web browser containing new features, such as icons as links. In 1994, Andreeson and his team left to set up a company that would produce a commercial version of Mosaic called Netscape. With information now easy and fun to access on the web, the Internet became a household word.[52]

To connect sources and users that have disparate characteristics is the basic idea behind internetworking. The Internet increasingly connects wired telephone systems, cellular systems, satellites covering remote areas inaccessible to wires or cell stations, and others. By integrating communication, computation, and microelectronics, engineers have developed a multimedia internet that carries voice, video, and computer data. Further into the future, they envision a global information-tion highway that will enable communication with anyone, anytime, anywhere, in any form, moving at any speed, and using any compatible terminal.

4

ENGINEERS IN SOCIETY

4.1 Social Ascent and Images of Engineers

Engineers pride themselves on solving real-world problems and getting things done creatively, scientifically, effectively. Practicality and efficiency, however, are not universally valued. "The idea of utility has long borne the stamp of vulgarity," a historian remarked on the literary culture, where the concept of technology failed to gain currency until after World War I and even now is deemed a "hazardous concept."[1] Differences between the cultures of doers and talkers spawn misunderstanding. Fortunately, stereotypes of engineers have not eclipsed their socioeconomic contributions.

Historically, engineers and their predecessors came mostly from working families, toiled with their hands, relied more on their own thinking and experience than on schooling, and were obliged to deliver products on demand. They were looked down on by ladies and gentlemen of letters and leisure who fancied themselves cultured and superior to those who produced the goods and services that made possible their elegance and opulence. Elitist slings and arrows were felt by Leonardo da Vinci, who replied: "I am fully conscious that, not being a literary man, certain presumptuous persons will think that they may reasonably blame me; alleging that I am not a man of letters. Foolish folks! do they not know that I might retort as Marius did to the Roman Patricians by saying: That they, who deck themselves out in the labours of others will not allow me my own."[2]

Before he entered government, Herbert Hoover visited England and joined a debate to argue that Oxford and Cambridge universities should include engineering in their instruction. On the Atlantic passage back to America he shared a dining table with an English lady and they had many delightful conversations. At the farewell breakfast she asked his profession. Upon hearing that he was an engineer, she emitted an involuntary exclamation: "Why, I thought you were a gentleman!"[3]

Engineers surprised not only English ladies. Theodore Roosevelt, U.S. president, wrote to a friend about John Stevens, whom he appointed as the chief engineer to build the Panama canal. That engineer, he discovered, read a lot of literary books. And he read not only novels but also poetry. "This I regard as astounding," Roosevelt wrote.[4]

Triumph of Self-Education

Stevens came from the backwoods and educated himself while working on the railway. This was almost the norm for early engineers. Most technical pioneers of the industrial revolution came from the lower social classes and had little schooling. Even among leading engineers, only a few had any higher education. Some had attended grammar school or high school. Others had spent their childhood working in mines or at arsenals. The majority had been apprenticed to millwrights, barbers, carpenters, stone masons, instrument makers. Because of their humble origins, these pioneering engineers were often stereotyped as craftsmen incapable of scientific reasoning. That stereotype persists. As late as 1959, when engineering research was widespread, a scholar declared that no word existed for an improver of technology comparable to "scientist" as an improver of knowledge. He rejected "engineer" because "it originally, and essentially still, refers to the expert practitioner of one of the arts."[5]

That stereotype is wrong even for early pioneers. We saw the scientific thinking of Smeaton and Watt, in section 2.3. Painstaking historical research has shown that although the leading engineers had not received formal education, they were not uneducated. As historians A. E. Musson and Eric Robinson remarked, the erroneous stereotype of early engineers as illiterate and unscientific tinkers is "a direct outcome of the difficulty which academically educated people often have in understanding the possibilities of self-education."[6]

Educate themselves pioneering engineers did. In Britain young Fairbairn burned the midnight oil to learn mathematics, Smeaton limited his business to find time for scientific studies, and George Stephenson had his son Robert borrow technical books from the library and read them aloud to him in the evening. Telford wrote in a letter: "Knowledge is my most ardent pursuit," and despite his preference for the experimental approach, he read widely and collected a huge library that included many French engineering and theoretical books.[7] Several engineers, including Smeaton, Maudslay, Fairbairn, and Nasmyth, were interested in astronomy. They made their own telescopes and observation instruments, thus demonstrating not only the requisite mathematical competence but also a broad intellectual curiosity. The spirit of self-help leaps from the pages of Samuel Smile's *Life of Engineers*. With little if any formal training, they transformed themselves from artisans to engineers and sometimes successful entrepreneurs. Often one man excelled in both civil and mechanical engineering. The Stephensons built not only steam engines and locomotives but also the railways and the bridges that they ran on. Smeaton not only built the Eddystone Lighthouse but also performed systematic experiments to improve efficiencies of water wheels and reported his results in scientific papers. Examples multiply, and they are not limited to the British. American engineers such as Edison readily come to mind. With their independence of thought and keen intuition, these autodidacts absorbed the relevant scientific knowledge and integrated it with the practical know-how acquired through their work, thus elaborating it and creating a new body of engineering knowledge. As historian Sidney Pollard remarked, "They rose from the ranks, so that in one lifetime they had not only to make the transition from rags to riches, but also to absorb, and indeed create, a completely new technology."[8]

Efficacy of Informal Association

Self-education requires motivation and access to existing knowledge. It is unclear why in that particular time and society so many people were so motivated, but opportunities were present in the British economic expansion. In response to demand, the eighteenth century saw the publication of a profuse number of textbooks and periodicals, and other printed matter on mathematical, scientific, and technical topics addressed to the popular audience. Lectures disseminating scientific knowledge, popularized in London, were later spread around the country by itinerant lecturers. The fascination they exerted on ordinary people was captured by Joseph Wright in his paintings, *A Philosopher Giving a Lecture on the Orrery* and *An Experiment on a Bird in the Air Pump.*[9]

To working men with high motivation and ability, the doors of the scientific community were not closed. Associations were not formal but, contrary to stereotypes, they were not in vain. Many conference goers think they reap the most benefit not from formal talks but from informal chats in the corridors. A formal theory has its own systematic structures, most of which may be irrelevant to the problem at hand, and learning them may be a distraction and a waste of intellectual resources. In informal conversation between two knowledgeable persons, one can quickly home in on relevant pieces of information. Broad visions, open-mindedness, critical analysis, and other rational ways of thinking crucial to science are also more likely to be picked up through personal contacts than in formal theories. Such informal interactions occurred during the industrial revolution. Many clubs and associations existed in which philosophers, scientists, engineers, and industrialists mingled to discuss their mutual interests. Among them was the Lunar Society of Birmingham, which met monthly when the moon was full throughout the late years of the eighteenth century and into the nineteenth. Its members at one time or another included Smeaton; Watt and his partner Matthew Boulton; William Herschel the astronomer; Joseph Priestley the chemist who isolated oxygen; Josiah Wedgwood, an industrialist steeped in chemistry; and Erasmus Darwin, a physician and the grandfather of Charles. They brought visitors, demonstrated experiments, asked questions, and gave advice, thus providing mutual support as they pursued their manifold interests. Such was the scientific ground in which the seeds of technological ideas sprouted and grew.

The prevalence and overall importance of interaction between scientists and engineers should not be overstated. It nurtured leaders, but a long road still lay between craft and scientific engineering. Even knowledge that would have been readily applicable and profitable in mechanical crafts, such as the optimal shapes for teeth of gears as worked out by mathematicians, waited in oblivion for decades. Nevertheless, some interaction did occur. The ability to discuss issues as equals with leading scientists showed that many engineers were anything but incapable of scientific thinking.

Establishment of Professional Associations

More important than mixing with scientists, engineers formed associations among themselves to exchange information and discuss problems peculiar to their concerns. Smeaton initiated the Society of Civil Engineers in 1771, which was little more than a dining club at first. The British Institution of Civil Engineers (ICE), established in 1818, was the first professional society in the modern sense. Its president, Telford, exhorted members to volunteer their effort to keep it going. Unlike their French colleagues, they had no government sponsorship, but then they were free from government interference. From their craft origins, autodidactic British engineers rose to enunciate the vision of a new profession in the charter of the world's first engineering society: "Engineering is the art of directing the great sources of power in Nature to the use and convenience of man; being that practical application of the most important principles of natural philosophy which has, in a considerable degree, realized the anticipations of Bacon, and changed the aspect and state of affairs in the whole world."[10]

The fledging ICE acquired its own premises and library, organized weekly meetings, and recorded the proceedings. Soon it started to publish engineering journals, coordinate the dissemination of technical knowledge, establish rules for membership, and standardize certain measures and practices. It served as the model for other societies. Unhappy about ICE's overemphasis on construction, mechanical engineers established the Institution of Mechanical Engineers in 1846 with George Stephenson as the first president. Engineers in other areas followed suit.

The German Society of Engineers, formed in 1856, played a crucial role in bringing German polytechnic schools first up to university status and then into pioneering graduate education. The American Society of Civil Engineers appeared in 1852. Societies proliferated because of the diversity of engineering specialties. Efforts to form a national fraternity of engineers started in 1886, culminating in the foundation of the U.S. National Academy of Engineering in 1964.

Rise of Engineering Universities

Professional societies organize active workers, and their journal articles are mainly addressed to knowledgeable readers. As with any human enterprise, a profession must nurture its future generations. A knowledge-intensive profession such as engineering further requires the organization and presentation of knowledge in forms accessible and stimulating to the novice. Self-education and informal learning can go a long way, but they are too unpredictable to produce a sufficient number of qualified engineers, especially when technological complexity gallops ahead. Formal education connects generations of workers more consistently. In this Britain lagged. Its major universities, engrossed in classics and the liberal arts, provided technical education belatedly and reluctantly. As the industrial revolution progressed, it began to lose its technological leadership. An engineer said at an 1893 ICE meeting: "The danger which, as many thought, now threatened English engineering, lay in the more thorough education and superior mathematical knowledge of so many foreign engineers."[11] The foundation of that threat, laid centuries earlier, revealed the role of government as a partner in technology.[12]

Practical artists who undertook public works and military projects had a closer relationship with governments than workers in many other trades. Recognizing their importance and wanting to protect them from the arrogance of soldiers, the French government created the Corps des Ingénieurs du Génie Militaire in 1675. Like today's U.S. Army Corps of Engineers, French officers in the corps often engaged in constructing roads, bridges, and other civilian works. It was at this time that "engineer" first became a title for professionals with special expertise. To educate them better, the government established around 1747 several military academies equipped with first-rate physics and chemistry laboratories. These schools offered rigorous curricula that included cartography, machine design, building construction, and especially mathematics. One of their graduates, Coulomb, became a renowned engineer and then physicist.

The greatest change occurred when the French Revolution unleashed enthusiasm to improve the human condition by rational means. The École Polytechnique, created in 1794, was an educational breakthrough not only in engineering but in science generally. Some four hundred students, selected by competitive examination from all over France, enrolled in the first year. They started with two years of scientific instruction that included calculus, mechanics, and chemistry, so that they had a grasp of general principles before they specialized in various branches of engineering. Among their professors were the best engineers, mathematicians, and natural scientists, and many of the students embarked on equally brilliant careers. Graduates from the Polytechnique secured the reputation of scientific engineers. Napoleon would not let them into battle even in dire crises, saying he would not wish to kill the goose that lays golden eggs.

Sponsored by the government, the École Polytechnique was subject to the vicissitudes of politics, which were especially violent in France at that time. It was also criticized for being too theoretical. Several private technical colleges appeared, offering more readily useful knowledge to a greater number of people. Chief among these was the École Centrale des Arts et Manufactures, which boasted Eiffel and Jenny as alumni.

The French model was emulated worldwide. In America, the military academy at West Point, first established in 1802, was reorganized in 1812 with the help of French-educated officers who introduced an engineering curriculum along the line of that at the École Polytechnique. It produced far fewer engineers than its French counterpart but shared its vulnerability to competition from the private sector. Rensselaer Polytechnic Institute, patterned after the École Centrale, appeared in 1849 and became the model for several others. Among its first graduates was Washington Roebling, who succeeded his father, John, in building New York's Brooklyn Bridge. Demand for courses outpaced supply, and thus construction sites and industrial shops continued to be major spawning grounds for engineers. Technical education in the United States received a big boost from the Morrill Land Grant College Act of 1862, which granted federal lands to institutions of higher education with the stipulation that they offer instruction in "agricultural and the mechanic arts." Among others, it aided MIT, Rutgers, Yale, and the University of Michigan, and it created Cornell,

Purdue, and the state universities of Ohio and Iowa. Ten years later the nation had seventy engineering colleges.

Enrollment in engineering courses increased much faster than in liberal arts courses. To accommodate the torrent of students, in 1895 the University of Michigan created an independent engineering school, whose enrollment grew from 331 to 1,165 during the next decade while its school of literature added just over 300 students. Similar enrollment patterns were seen at other universities. Together, universities awarded 226 engineering degrees in 1880 and some 11,000 in 1930. And the trend continued. Engineering soon became a favorite subject of college students. (For more recent profiles, see Appendix A).

Social and Sociological Images of Engineers

Besides leading technological revolutions, engineering institutes contributed to the democratization of education. By lightening up the prevailing requirements for arcane classics and replacing them with modern subjects useful in productive occupations, they opened a route to upward social mobility for children of the lower and middle classes who hitherto could not afford higher education. The social difference between the new students and old elites was noticed early on. MIT's president reported in 1894: "An unusually large proportion of our students come from families of moderate means [who] never consider the possibility of first going to a classical college and then to a professional school." Farther up the Charles River, Harvard's president observed marked differences between liberal arts students and those in engineering—"young men who know from the start the profession they are destined for."[13] The techies were coming. Talented, motivated, determined to contribute to and benefit from fast-moving industries that rewarded productive achievers, they quickly proved that workers dedicated to producing socially beneficial goods and services could excel not only with their hands but also with their brains. The old cultural aristocrats who scorned them were soon discomfited by the emergence of a rival culture.

Folks cheered the new professionals who lit up their homes with electricity, connected them by telephone, brought freedom of mobility with cars and bridges. In an extensive study of popular culture, historian Cecelia Tichi found that "the engineer became a vital American symbol between the 1890s and 1920s. In popular fiction, as we shall see, he signified stability in a changing world. He was technology's human face, providing reassurance that the world of gears and girdles combined rationality with humanity." In this period, the engineer appeared as the hero in more than a hundred silent movies and in best-selling novels that sold nearly five million copies. "In these popular novels the engineer's vision is both prophetic, in that it foresees the future, and poetic, since it originates within the realm of imagination . . . These engineers realize their dreams. They bring the bold, visionary schemes from the imagination into tangible reality. They are poets for the quotidian."[14]

The young profession received a totally different treatment from the old establishment. One historian cited an engineer who in the late 1890s predicted that because electricity would revolutionize society, engineers would be the "priests of the new epoch." If so, the historian asked, "why are engineers so invisible in American culture? Is it because, after all, they are not synonymous with technology? Or is it that, in any event, elevated literature is the wrong place to look for them?" This elitist reduction of culture to elevated literature I will discuss shortly. Such disdain was corroborated by Tichi, who observed that the engineer was "the invisible man in American studies." Doctors, lawyers, and other professionals have been discussed in college classrooms and scholarly journals, but not engineers; "critics and analysts have been oddly blind to their presence."[15]

When technological achievements could no longer be ignored, blanket contempt changed to a partial condescension expressed in two stereotypes, which have been extended to natural scientists when their research results have proved to be of practical use and thus less than "pure." The first is *Dr. Frankenstein,* who makes monsters. The second is the *nerd* or the *geek:* bright and proficient but bloodless and deficient in social relationships. The stereotypical transition from the illiterate artisan to the "antisocial brainiac" was described by a historian in discussing the term "institute of technology." He wrote: "This abstract word [technology], with its vivid blankness, its lack of a specific artifactual, tangible, sensuous referent, its aura of sanitized, bloodless cerebration and precision, helped to ease the introduction of the practical arts—especially the new engineering profession—into the precincts of higher learning."[16] Compare this picture of technology to Aristotle's description of *téchnē* as the capacity to produce with a true *logos,* presented in section 2.1, and one sees how much the literary culture has changed.

Cultural resentment increased after World War II, when technical professionals became socially influential. Engineers were referred to as "narrow gauge" and "domesticated servants of capital."[17] Surveying the scholarly literature, another historian observed: "Ever since John Alexander Law Waddell started telling engineering students to clean their fingernails and comb their hair, it has been easy to caricature the profession . . . Instead of the portrait of a profession, what we have is a grab bag of stereotypical images and they picture a group that seems politically inflexible, socially awkward, culturally limited, and ethically inert."[18]

Are the uncouth stereotypes of engineers justified? We glance at some general rebuttals here; many concrete examples are presented in Chapter 7. William Rogers, founder of MIT, wrote in his proposal for the new institute: "In the features of the plan here sketched, it will be apparent that the education we seek to provide, although eminently practical in its aims, has no affinity with that instruction in mere empirical routine which has sometimes been vaunted as the proper education for those who are to engage in industries. We believe, on the contrary, that the most truly practical education, even in an industrial point of view, is one founded on a thorough knowledge of scientific laws and principles, and one which unites with habits of close observation and exact reasoning, a *large general cultivation.*"[19]

Ever since MIT opened its doors, it has required students to take history, literature, and other humanities courses besides science and engineering. This practice is common in engineering education. The advice that schools should make economics and the humanities required courses, issued by the Society for Promotion of Engineering Education in the 1920s after a multiyear study, had a profound influence. A 1955 report for the American Society for Engineering Education proposed that about a quarter of undergraduate courses be devoted to humanities and social sciences, and the remaining three quarters divided among the relevant natural sciences and engineering specialties.[20] This mixture has become almost the norm in American universities, as the 1955 report formed the basis for professional accreditation of engineering curricula. Instruction in engineering usually does not start until the sophomore year. Freshman engineering majors compete head-on with history majors in required history classes, and they are not offered "English for engineers" classes akin to the "physics for poets" classes often offered for nontechnical students. Some universities, for example Stanford, demand more than the general university humanities requirements for engineering majors, who must take additional courses in the relationship between technology and society. Generally, engineering and natural science majors take more humanities classes than liberal arts majors do science and technology courses. On the Graduate Record Examination for entrance to graduate school, engineers and natural scientists consistently and comfortably beat social and behavioral scientists in *verbal* scores, not to mention analytic and quantitative skills.[21]

The nerd stereotype for engineers harps on their technical language, which is incomprehensible to the layperson. However, words must be judged with respect to their meanings and audiences; often difficulty lies not in the language but in the topic discussed. Some topics are more complex and unfamiliar than others to the general populace. People know this; as they sometimes protest about opaque explanations that they think should not be so difficult: it is not rocket science. Modern technologies are often radically new and tremendously complicated, and to make them work engineers must be precise in their thinking and speech. Analysis introduces distinctions, which require new concepts and words or symbols for representation. Therefore it is no surprise that engineers develop technical terminologies. And they are not the only ones with jargon. In this the literary culture is not to be outdone. Feynman had experience with both. As a commissioner investigating the space shuttle *Challenger* disaster, he talked extensively with rocket engineers. Communication went very slowly at first; he faced a huge amount of information filled with unfamiliar technical jargon on complicated rocket technologies he was ignorant of. Then he caught on and understood the engineers: "I gained a great deal of respect for them. They are all very straight, and everything was great." Compare this with his experience at a conference intended to bring together scientists and scholars. It too began slowly; the sociological papers presented were totally opaque to him. Frustrated, he picked out one long, jargon-filled sentence and read it over and over again until he got its meaning: sometimes people read. "Then it became a kind of empty business: 'Sometimes people read, sometimes people listen to the radio,' and so on, but written in such a fancy way that I couldn't understand it at first, and when I finally deciphered it, there was nothing to it."[22] Sometimes jargon expresses complex contents, other times it disguises trivial contents. It is ironic that the former are more often stereotyped as nerdy.

From Two Cultures to Science Wars

Eric Ashby, president of Queen's University in Belfast, observed in 1959 the conflict between the cultures of "the scholar-gentleman" and "the practical man." He wrote, "The crude engineer, the mere technologist (the very adjectives are symptoms of the attitude) are tolerated in universities because the State and industry are willing to support them. Tolerated, but not assimilated; for the traditional don is not yet willing to admit that technologists may have anything intrinsic to contribute to academic life."[23] Within a year, the gap between the "literary culture" and "scientific culture" was brought to public consciousness by scientist and novelist C. P. Snow, who observed that the literary culture discriminated against not merely practicality but science itself; its definition of "intellectuals" excluded not only engineers but also mathematicians and pure scientists.[24] These observations are anything but dated today. Snow's description of the literary culture is remarkably apt to what Harvard University president Lawrence Summers criticized in 2002: "a culture where it is unacceptable not to be able to name five plays by Shakespeare but where it is fine to not know the difference between a gene and chromosome."[25]

Snow argued that those ignorant of the second law of thermodynamics are as one-sided as those ignorant of Shakespeare; the literary culture does not have a monopoly on the humanistic spirit; scientists have a culture that is just as rich in humanism, if not more so. His argument provoked battle cries from those on the literary side eager to protect their turf. Insisting that scientists and engineers are in need of being "acculturated" and "humanized," some literati advanced ideologies aiming to take on and deconstruct science and technology: postmodernism, cultural relativism, the strong program, the sociology of scientific knowledge, social constructionism. Postmodern ideologues mock rationality, deride references to the real world, deny the objectivity of science, and attack accounts by scientists and engineers of their own works as "myth" and "Whiggish." They have invented what they call "the scientist's account," depicting scientific discovery as a mechanical activity that simply receives input from nature and claims absolute certainty for not involving human judgment. They use this caricature of the scientist as a whipping boy in promoting their own doctrine that all scientific facts, including physical laws, are thorough social constructions "true" only in the culture in which they are constructed. Since the late 1990s, many scientists and engineers have spoken out and criticized these practices for their conceptual confusion, factual distortion, and misrepresentation of science and technology. They organized a conference to criticize what they deemed "the flight from science and reason." A leading postmodern journal countered with a special issue entitled *Science Wars*. The dispute has broadened, accumulating a large literature and making more people aware of the festering resentment that needs to be addressed.[26]

In 1959 engineers joined scholars to found the Society for History of Technology (SHOT). In 2001 historian Terry Reynolds, in his presidential address to the society, appealed: "The engineering community played a major role in SHOT's emergence, and while we no longer look to that community as our primary patron and audience, we should not deliberately burn the remaining bridges that link our field to theirs." He described how authors who attend to engineering and scientific contents are put down as "rivet counters" and made to feel inferior. Their works are snubbed as being "passé," "hardware catalogues," and "nuts-and-bolts." "Hardware historians need not apply" flashes blatantly in some job announcements.[27] Economics is put down alongside engineering and natural science as being merely nuts-and-bolts. The repression of hardheaded topics by fashionable ideologies is a hot spot in the "science wars" that afflicts areas far beyond SHOT, to the dismay of conscientious scholars who find themselves marginalized. One lamented: "Scientists and engineers are not—or are no longer—welcome."[28] Another criticized her postmodern colleagues: "A certain 'Besserwisser' approach prevails, with the sociologists smugly overruling the scientists. It was as if the sociologists were the self-appointed psychoanalysts of scientists, knowing their 'true' motives, unbeknownst to the scientists themselves."[29] A third admitted: "We tend to lull ourselves into the conceit that we need only satisfy each other."[30] A fourth appealed to colleagues, that works "continue to be evaluated on scholarly merit and not on the basis of conformity to or rejection of any particular theory of history." Quoting that plea, Reynolds emphasized "the need for respect and toleration of seemingly obsolete approaches to scholarship in the history of technology, because these approaches may offer better bridges to certain of our external audiences than our most avant-garde scholarship."[31]

Technology now pervades our daily life and social infrastructure. Without some technical knowledge, it would be difficult to make responsible decisions in many practical situations, such as deciding whether to support fuel-efficiency regulations. People realize this; a survey found most Americans said they would like to understand technology.[32] Unfortunately, the spread of knowledge continues to be perverted by caricatures and intolerance. A college-level reader entitled *Major Problems in the History of American Technology* deliberately excludes engineering.[33] In a book that claims to survey historiography of technology across two decades, the ten-page index contains only one entry beginning with *eng-*: engineering theory.[34] The summary of a workshop on "convergent themes in technology" lists five themes: "institutions and systems, national and regional styles, inertia, ideology and rhetoric, and policy." Conspicuously absent are engineering and science; technology here is dominated by rhetoric, ideology, and cultural style.[35] Disturbed by this academic fashion, Columbia University provost Jonathan Cole collected the best books on American history. Combing through them, he found only trace mentions of technology, and those were mostly negative. Referring to Snow's analysis of the cultural gap, he concluded: "The young members of our society who become lawyers, businessmen, or members of Congress are surely not learning about the scientific and technology revolution in their reading of the best texts produced by American historians."[36] These texts also educate tomorrow's teachers and journalists, who will continue to spread the bias in schools and news media.

MIT president Charles Vest told the American Association for the

Advancement of Science (AAAS) in 2000: "Another major message, which I hope our Tech Forum can help spread in Congress, is that technology means more than just information technology."[37] Back at MIT, the director of the Science, Technology, and Society program alleged: "Inscribed on the Great Dome, the word 'technology' defines the mission of the institute. In everyday meetings held under the dome, the compass of the term has been drastically reduced, as already noted, to computers." Besides this allegation, made amid MIT's booming research activity in biotechnology, the director claimed: "While the boundaries of the engineering profession have expanded dramatically, as a profession it has lost its self-conscious moral stance . . . And so it is striking that in discussions about technological change at MIT, 'engineering' seems disconnected both from 'technology' and from 'progress.'"[38] When such a disconnect occurs within a leading institute of technology, imagine the message that reaches the public and Congress.

M. R. Greenwood observed in his 1999 presidential address to the AAAS that although most Americans are enthusiastic about science, American school systems are "facing a powerful and vocal minority of parents and community leaders who are distrustful of, and antagonistic toward, science." In a world where scientists and engineers are increasingly under the burden to explain science and technology to the public, education is too important to be left to antagonists across the cultural gap. Some engineering schools, perhaps in collaboration with schools of management and other practical professions, are gearing up to take the task upon themselves. If they do not work harder in this respect, science and engineering, with all their achievements, may be in danger of what Greenwood described: "Winning the battle but losing the war."[39]

4.2 Partnership in Research and Development

World population was expanding fast at the beginning of the twentieth century, and the supply of natural fertilizer was dwindling just as fast. Farmers badly needed artificial fertilizer to grow enough food for the multiplying mouths. In 1906 Fritz Haber, a physical chemist at Karlsruhe Technical Institute in Germany, discovered a process for synthesizing ammonia from hydrogen and nitrogen. It made possible the conversion of atmospheric nitrogen into forms useful for many things, including fertilizer. The process had great potential but seemed impractical, because it required pressures of up to 1000 atmospheres and temperatures up to 500°C. BASF, which had partially funded Haber's research and decided it warranted major investment, licensed the process and unleashed its in-house R&D staff. Carl Bosch, a mechanical engineer versed in chemistry, headed a team to scale up the process for production. They solved tremendous problems, including the corrosion by hydrogen of the high-strength steel required for high-pressure containers. The technology they developed commercialized synthetic fertilizer. Furthermore, as the world's first continuous industrial catalytic process, it opened the way to producing synthetic methanol and influenced oil refining.[40]

Besides being a breakthrough in science and technology, the Haber-Bosch process signified the triumph of organizational innovations: the research university and graduate school, the industrial laboratory, and their collaboration. The advent of organized R&D is sometimes called the invention of the method of invention. Its success required not only garnering talent but also nurturing it. BASF's laboratory, established in 1877, was the world's first industrial R&D facility. Success of such laboratories depends on the availability of capable research staff, which only universities can produce in quantity. Conversely, the ability of graduate schools to recruit top students depends on the availability of satisfactory R&D jobs. It is no surprise that graduate schools and industrial R&D had the same birthplace. That place, however, is somewhat surprising. It was Europe's backwater.

Beginning of Graduate Education

Before unification in 1871, Germany was an idea that did not exist in reality. When Paris and London held the torches of culture and commerce, the German states were backward provinces. Determined to catch up, the Germans embraced the Enlightenment enthusiastically and developed an idealist and cosmopolitan philosophy in the eighteenth century. Maintaining that the unknown remained vast, they rejected theories that regarded education as the mere transmission of existing knowledge. Instead, they insisted that education should cultivate the capability for independent and critical thinking in young people, so that they could go on to search for new knowledge.[41]

Scientists had performed research for centuries, but there was scant organization for it, nor was there any systematic effort to induce students into it. Research was mainly an individual endeavor initiated by personal aspiration; financed haphazardly or even by personal means; promulgated by irregular publications, informal correspondence among like-minded people, and visits to working laboratories. Just like self-education, self-training in research can produce individual geniuses but not a corps of qualified researchers. Realizing this, German universities began to organize for serious scholarship and research. Supported by state authorities, they were the first to award Ph.D. degrees, foremost to physicists and chemists.

An outstanding example of pedagogy in a research university was Justus Liebig's chemistry laboratory at the University of Giessen, which from architecture up was designed for research. Besides its many discoveries in chemistry, it pioneered a new mode of education that engaged students in research. Students spent six months a year in preparatory courses. The other six months they worked in laboratories, interacting closely with the faculty, acquiring a wide view of a yet unexplored area, and developing the ability to think independently and creatively. Graduate education was a genuine novelty in the early nineteenth century. Students from all over Europe flocked to Giessen. Through the lineage of Augustus Hoffmann and William Perkin, the university sowed the seeds for the dye industry that would blossom decades later.

German universities were not exempted from a snobbishness toward engineering and practical arts. These fell into the domain of technical institutes, which appeared in Karlsruhe (Baden) in 1825 and Berlin in 1827. Poor images of the École Polytechnique at first, they worked hard to improve, acquiring university status by the 1860s and the right to award master and doctorate degrees in engineering by the end of the century. During the 1900s, Germany's eleven technical institutes and three mining colleges awarded almost thirteen hundred engineering doctorates and ten times as many masters. To this day, engineers enjoy a high social status in Germany.

American research universities had a slow start. After the Civil War, educators called for more emphasis on new fields in natural science and advanced studies similar to those found in Germany. By 1920, when Throop College became the California Institute of Technology, or Caltech, some twenty American universities performed research. They included colonial establishments such as Columbia, Harvard, Princeton, University of Pennsylvania, and Yale; state universities such as California, Illinois, Michigan, Minnesota, and Wisconsin; and private institutes of late-nineteenth-century vintage: Chicago, Cornell, Hopkins, MIT, and Stanford. Together they had about 340 research fellows and $1.8 million in research grants ($14.4 million in 1998 dollars). Most of the money was put up by philanthropic foundations, the bulk earmarked for medicine.

Engineering programs lagged. Engineering research laboratories appeared at the University of Illinois and MIT in 1903 and were operated at thirty-eight universities by 1937. Their scales were modest, with a total budget of $264,000 in 1920 and $11 million in 1937 (both in 1998 dollars). Engineering schools had difficulty attracting graduate students, as their baccalaureates were snatched up by the booming electrical industry. The nation's top four universities cumulatively had awarded only ninety-two masters and eleven doctorates in electrical engineering up until 1919.

Fundamental Principles and Industrial Applications

Because universities aim to nurture analytic and problem-solving abilities together with imparting reasonably broad technical knowledge, the content of their curricula does not always dovetail with the needs of specific industries. This potential mismatch has been a problem since the very beginning of undergraduate education, and the situation was worse for graduate schools. To address the problem, many universities established cooperative ties with industry. Such ties were crucial to the early success of German research universities. The universities appointed professors with industrial experience who were knowledgeable about cutting-edge technology. Companies hired fresh doctorates, had them work in manufacturing to learn about the real world, and then sent them to a technical institute to perform research related to industrial problems. This academic-industry relationship proved fruitful. Academic chemists at the Berlin Technical Institute synthesized alizarin, a dye superior for its color-fastness. BASF and Bayer quickly commercialized the process, setting the foundation for a strong German dye industry.

Americans followed a similar path. MIT established its School of Chemical Engineering Practice in 1916. It arranged to station faculty members in companies widely representative of the chemical industry. Students spent eight weeks each in three separate plants where, under guidance of the resident professors, they solved problems with real-world consequences. For the rest of the year, they returned to MIT for courses designed specially for them. Their internship in industry served them in lieu of a master's thesis. With variation in details, similar cooperative programs were set up at many engineering research universities, ensuring the practicality of education.

Cooperation is not the same as merging, however. Universities and industries have different missions, needs, and approaches, which sometimes conflict. A debate erupted when MIT developed its first research programs. Arthur Noyes stressed the importance of basic science: "What is wanted of the factory chemist in this country is rather the power of solving new problems and making improvements in processes —a power to be acquired far more by a good chemical training, which should include a large proportion of research and other work requiring independent thinking." William Walker disagreed: "It is a much smaller matter to both teach and learn pure science than it is to intelligently apply this science to the solution of problems as they arise in daily life." The two positions were not necessarily incompatible, although their proponents may have had a personality conflict. Their synthesis would later become the intellectual foundation from which Caltech rose meteorically and MIT made itself into a world-class research university. At the time, though, Caltech was not on the map and MIT was a struggling local school.[42]

Noyes in 1903 dipped into his own pocket to finance the research laboratory of physical chemistry that produced MIT's first Ph.D.'s. Five years later Walker set up a rival research laboratory of applied chemistry, which derived its operating budget mostly from contracts with industrial firms. It also served as the cornerstone for a chemical engineering graduate program. In view of its popularity among students and industrialists, MIT in 1920 put Walker in charge of its new Division of

Industrial Cooperation and Research, which ran a program to do research for industries.

The applied laboratory was prosperous, but something went wrong. Conceived to perform broad research applicable to a wide range of industrial processes, it was increasingly obliged to solve narrow problems for its client firms. Some tasks, for instance finding a way to manufacture leak-proof oil barrels or waterproof paper, hardly qualified as research. They were closer to what William Wickenden, a former MIT professor who chaired a big project to evaluate engineering education nationwide, described in 1929 as "first-aid problems for industry, routine engineering testing, and what may be called quasi-scientific puttering."[43] The ranks of faculty revolted against what they regarded as cash cows for consulting firms rather than research projects for universities. Noyes headed west, followed by his best students, and helped to turn Caltech into a rival with a surpassing scientific reputation. Walker left for his own consulting business. MIT failed to attract research fellows or raise research funds from philanthropic foundations generous to other universities.

Alarmed by its sinking prestige, MIT in 1930 appointed Karl Compton, a physicist, as its new president. To restore balance and reverse the slide, Compton steered the institute back toward basic science without forsaking applied research. This he achieved by preserving the valuable ties with industry that Walker had built up but regulating them in conformity with the mission and interests of a science-based research university. The Division of Industrial Cooperation evolved into today's Industrial Liaison Program, a widely emulated model for effective technology transfer from academic research to industrial innovation.[44]

The MIT episode is not a dusty story. Maintaining a proper balance between basic research and its applications is a perennial problem, continually discussed by engineering educators as they modify curricula to keep ahead of advancing science and technology. It is important to note that what hurt MIT's academic reputation in the late 1920s was not the *applied* nature of its research but the *degeneracy* of its research. Degeneration can occur in any kind of research, not least in those areas without any application in the real world. When the problem was corrected, MIT's applied research continued to thrive under Compton. What Noyes promoted was not the *purity* of science but the *mastery of fun-*

damental scientific principles in applications; he was a leader in organizing the application of science to assist America's effort in World War I. And when it came down to deeds instead of words, Walker did not denigrate fundamental principles either. Developing principles for chemical engineering was precisely what he strove for.

Engineers' attention to science is not vanity or social fashion, as claimed by scholars who snub technical content. It is the result of realistic judgments about the jobs at hand and to come, the human resources they require, and the effective division of labor between university and industry. Noyes was one in a long line of educators who saw this. Navier and other civil engineers developed mathematical structural analysis and taught it at universities. British engineering professor William Rankine, who used thermodynamics to explain operations of steam engines and facilitate designs of prime movers, had elaborated the notion of *engineering science* in the 1850s. His approach to engineering was promoted in America by Robert Thurston, who developed the mechanical engineering curricula for Stevens Institute of Technology and Cornell. On the industrial side, Steinmetz, General Electric's chief engineer and the designer of many alternate-current devices, argued that universities should be responsible for producing students with adequate mastery of the basic science required for industrial engineering. Wickenden, AT&T's vice president, arrived at a similar conclusion in his report on engineering education. GE president Gerard Swope and Bell Labs director Frank Jewett were among those who in 1930 advised MIT to turn to fundamental science and integrate it with engineering.[45] In short, academic and industrial leaders agreed that technology had advanced to such a degree of complexity that engineers could not succeed without certain scientific capabilities. *Theoretical* does not mean impractical. Quite the contrary, some theoretical ability is required for adaptive and innovative engineering. The main mission of research universities as educational institutions is not so much to produce solutions to specific problems as to nurture problem solvers able to take on a wide range of challenges.

Maturation of Engineering Graduate Education

"What is an engineering research Ph.D. for?" asked John Armstrong, IBM's vice president of science and technology in the late twentieth century. He answered that the most valuable edge a Ph.D. holder has is knowing "how to learn at a very sophisticated level" and "how to approach and solve problems starting from powerful and fundamental points."[46] The answer reveals a deep awareness of the uncertainties in the real world, which can at any time spring undreamed-of problems whose solutions demand abilities that are not only sophisticated but also versatile, an adaptability acquired by mastery of fundamentals. It was such an unexpected situation that gave a decisive impetus to engineering education in America.

World War II called on engineers and scientists to develop radar and other high-tech weaponry with lightning speed. The demanding tasks revealed their differences. Terman, who directed the Radio Research Laboratory at Harvard, recalled his difficulty in finding engineers up to the job: "With the beginning of World War II, electrical engineers with a four-year background augmented by practical experience were found to be wanting in the exciting new areas of electrical engineering involving microwaves, pulse technology, computers, diodes, etc. Instead, Ph.D. physicists who knew very little conventional electronics and no engineering, but whose education included three or four years of graduate work and a sound background in classical physics and mathematics, took over, almost to the exclusion of electrical engineers."[47]

"Never again will electrical engineers be caught short," vowed Terman after the war. His vow was shared by leading engineers in other fields who had had a similar experience in their war efforts. Together they made it good by revamping engineering curricula and putting engineering sciences on a firm foundation. Terman promoted an undergraduate curriculum at Stanford that emphasized the fundamental scientific principles and a graduate program that nurtured original thinking in student-faculty research. Equally adamant about engineering practicality and the university-industry connection, he initiated a program that made the Stanford Research Park a spawning ground for Silicon Valley. His approach to engineering education exerted wide influence.[48]

Offered exciting programs, fueled by ample research funds, and incited by employment opportunities, graduate engineering education flourished. In 1946 only 5 percent of all master's degrees awarded in America were in engineering; four years later the number had jumped to 8 percent, and it continued to rise. Today the number of masters in engineering awarded each year is only slightly less than half of the number of bachelor's degrees. The number of engineering doctorates increased less rapidly but steadily. In 2000, American universities awarded 5,430 doctorates in engineering. About three quarters of working doctoral engineers engage in R&D.

Research and Development

Leaders in technology look beyond *invention* and emphasize *innovation*. Innovation is a comprehensive process that includes stimulating and directing appropriate effort; inventing; bringing the invention through development and production to market, financial return, and national economic growth. When the goal is set at innovation, the boundary between research and development is blurred. Publication becomes not an end in itself but a means to make knowledge freely accessible for use. The mere desire to have discoveries developed and commercialized influences the direction of research, and the development of a product may raise new difficulties that call for further research. Because of their reciprocal relationship, the concepts of research and development were linked in the early 1920s and now R&D is said in one breath.

To facilitate, coordinate, and evaluate R&D, the National Science Foundation stated,

> The objective of basic research is to gain more comprehensive knowledge or understanding of the subject under study, without specific applications in mind. In industry, basic research is defined as research that advances scientific knowledge but does not have specific immediate commercial objectives, although it may be in the fields of present or potential commercial interest. Applied research is aimed at gaining the knowledge or understanding to meet a specific, recognized need. In industry, applied research includes investigations oriented to discovering new scientific knowledge that has specific commercial objectives with respect to products, processes, or services. Development is the systematic use of the knowledge or understanding gained from research directed

toward the production of useful materials, devices, systems, or methods, including the design and development of prototypes and processes.[49]

Development, which does *not* include routine product testing and quality control, is no less complex than research. For a specific product, it often consumes ten times the resources invested in research. Research and development have many differences and much in common. They venture into the unknown and undertake risks, guided by knowledge and educated guesses, although the guesses are more speculative for research than development. Compared with research, development is less risky, more regimented, has tighter time schedules, and is subject to more stringent budgetary controls. Unlike researchers, who often define their own problems, most developers receive problems and goals. Development has a narrower focus and more predictable results, but it is not therefore dull and unchallenging; ingenuity is directed more at overcoming the difficulties imposed by exacting constraints. Climbing Mount Everest is more regimented than exploring an unknown continent, but it is no less exciting.

Industrial R&D

Heinrich Caro, BASF's first research director, said that industrial research banks on "massive scientific teamwork" instead of the efforts of individual scientists. A genuine team is more than the sum of its parts, because its members cooperate. Willis Whitney, GE's first research director, explained: "Separated men trying their individual experiments contribute in proportion to their numbers, and their work may be called mathematically additive. The effect of a single piece of apparatus given to one man is also additive only, but when a group of men are *co*operating, as distinct from merely operating, their work rises with some higher power of the number than the first power. It approaches the square for two men and the cube for three."[50] The Englishman Perkin, whose chemical discovery initiated the dye industry, was brilliant. However, in the long run a single genius was no match for the exponential effects of organized research in the German industry.[51]

Industrial R&D has grown tremendously in America since GE set up its laboratory in 1900. Organization for research is less a method than an *enabling environment*. Industrial research laboratories do not have to be large establishments heavy with bureaucracy, or intellectual factories that stifle individual creativity and reduce invention to routine methodology. A corporation chooses the R&D organization appropriate for its products and business model. The success of many high-tech startups suggests that light brigades can be more effective than heavy cavalries, depending on the type of technology involved. And even in large laboratories, individual talent still matters a lot. Empirical studies have found that in corporations with significant R&D staffs, a tiny fraction of key inventors are responsible for the great majority of high-quality patents. Instead of blunting their ingenuity, R&D organizations facilitate it by providing a steady environment with adequate equipment and resources, freeing them from the business worries that overwhelm many inventor-entrepreneurs, and directing their attention to profitable areas. This does not mean that R&D staff members besides the key inventors are superfluous. A major patent contains an idea, which requires much technical development before it can be realized in a marketable product. Furthermore, a breakthrough has many ramifications that may also be exploited for maximum profit. For instance, once the basic mechanisms for synthesizing alizarin were known, BASF's laboratory proceeded systematically to synthesize new dyes by similar processes. Such jobs require substantial human resources, which R&D laboratories orchestrate.[52]

Research and development expenses cut into short-term profits. Commercial firms would not invest in R&D if their top management lacked long-term vision and the capacity to appreciate technological potential. Vision and technical ability are characteristics of engineers and scientists, and their leadership has been instrumental in the establishment of industrial R&D. The German chemical firms that pioneered industrial research were mostly founded by chemists and others knowledgeable in the technology. Most early champions for R&D in American corporate boardrooms, such as GE's Edwin Rice and AT&T's John Carty, were engineers and scientists. Some of the top executives who gave them full support, such as Alfred Sloan of GM and the du Pont family of DuPont, had similar backgrounds. Not surprisingly they also led in rational corporate structuring.

Leaders of industrial R&D share a general scientific outlook that fully appreciates the complexity of nature, the insecurity of fortune, the meagerness of human knowledge, and the remedial power of rational inquiry and planning. They realize that technologies and markets are dynamic and the future is uncertain. Therefore the capability to innovate and respond to unexpected change is crucial to the competitiveness and flexibility that ensures a corporation's long-term survival and prosperity. At the leading edge of technology, ignorance is usually great. A corporation cannot afford to wait for discoveries to be made by academic scientists or, worse, competitors. Even if a complacent company's core products are well protected by patents, it risks being outflanked and overrun by competitors capable of introducing superior alternatives and substitutes. In contrast, a vigilant company that through R&D has grasped the principles of its products has both offensive and defensive advantages. By exploiting the ramifications of the principles, they can discover new opportunities, introduce new product lines, and diversify into new areas, such as when a dye manufacturer diversifies into pharmaceuticals. Even if a company decides not to expand beyond its core business, it can still protect its flanks through appropriate patent acquisitions. The threat of Nernst glowers to Edison light bulbs was a strong motivation for R&D at GE, as the threat of wireless to wired telephony was for AT&T. As GE's Whitney put it, R&D is "life insurance" for its parent corporation.[53]

R&D stands as a unit in industry, and it is justified by its contributions to the parent company's competitiveness. Even AT&T's Bell Laboratories, with its sterling image as a research haven, was mostly a development organization that consistently devoted only about 10 percent of its budget to basic research. Xerox is often criticized for "fumbling the future" because its Palo Alto Research Center invented such devices as the laser printer and the computer graphic interface, but it failed to develop them and ruefully saw them exploited for great profit by other companies. When research centers grew fat, detached from development, and became irrelevant to corporate bottom lines during "the golden age of science" in the 1960s and 1970s, major reorientations followed. IBM Research, for instance, went through a restructuring so painful it was called a bloodbath. But cutting fat can be healthy. Armstrong said in the late 1990s: "It is startling. All the evidence—now this isn't easy for a former director of Research to say—is IBM Research is more effective now inside IBM than it's ever been."[54] Similar reorientations swept through corporate America. "Be vital to IBM" is an R&D strategy for IBM just as "Be vital to GE" is for GE. Basic research has not been forgotten; its share of total industrial R&D expenditures rose from 4 percent in 1980 to 9 percent in 2000. However, more than ever it must be "world class" research rather than "run-of-the-mill."[55]

R&D Services and Commoditization of Technology

The organization of industrial R&D changes with changing characteristics of technology and industrial structure. The rapid emergence of new technologies, the maturation of old ones, industry deregulation, and the shift in economic activity from manufacturing to service all contribute to reorganizations that extend far beyond individual firms. Traditionally, manufacturing firms such as GE and IBM have been the major homes of R&D laboratories, and they still are. Increasingly, however, large corporations are finding it more efficient to lessen their burden of peripheral technologies in order to concentrate on their core business. Thus they are downsizing their own laboratories and outsourcing parts of their R&D projects to consulting firms or university research centers. To perform these jobs, specialty firms are mushrooming. For example, the divestiture of AT&T broke up Bell Laboratories. The small science-oriented section of the old Bell Labs went to Lucent. The large engineering-oriented section finally became Telcordia, a leading consulting firm serving the whole telecommunications industry. In such restructuring, industrial R&D is gradually shifting from the manufacturing to the service sector.

Engineering services, in which engineers form specialized contracting firms and offer their expertise for a fee, have a long history. Some consulting companies perform R&D to improve their own technological expertise. Others perform R&D for outsiders on a contract basis. R&D in engineering services is a driving force behind the spread of new technologies and the technological evolutions essential to most mature industries. For example, many firms without in-house design facilities want new information technologies. To meet their demands, computer systems design services have become a huge business.

In 1997, engineering services in the United States, excluding those in architecture, brought in $88 billion. Engineers also participated in surveying services, testing laboratories, commercial R&D, and environmental and other technical consulting, which together billed $37 billion. On top of that, computer systems design and related services grossed $109 billion. The total came to $234 billion for engineering-related services, which almost doubled the legal-services total of $123 billion, although engineers are far less visible in society than lawyers.[56]

It may be that the same knowledge and technology are generated whether R&D is performed by manufacturing or service firms. Different industrial organizations, however, have a great impact on who has access to the knowledge and hence on technology diffusion. When a firm does its own R&D and keeps its technical expertise proprietary, it erects a technological barrier to other firms trying to enter the area. The barrier is lowered when R&D and technical expertise are offered as services available to all who can pay. Consulting fees are usually lower than the expense of maintaining an R&D facility. A company with no resources to develop its own technology base can nevertheless access the latest technology by contracting with the appropriate specialty firms. Thus engineering services contribute to the spread of technology. We saw in section 2.4 how engineering firms that developed turnkey chemical plants that anyone could buy contributed to the rapid rise of the petrochemical industry.

Another example of technological diffusion is in oil exploration and extraction. Twenty years ago the major oil companies kept most of the industry's technology proprietary, which deterred others who wished to explore new reserves. Now newcomers willing to invest in new reserves but with little technical expertise can hire geophysics surveying firms and drilling operators equipped with the best technology. Oil executive Peter Jones remarked: "To an increasing extent, technology has become a commodity."[57]

4.3 Contributions to Sectors of the Economy

In 2000, *Time* magazine called on its readers to name the most important event of the twentieth century. Some 400,000 votes were cast. Of the top twenty nominees, eleven were feats of science and engineering, led by the 1969 landing on the moon.[58]

Many organizations besides *Time* looked into the rearview mirror at the century's turn. The National Academy of Engineering asked professional societies to select twenty technological achievements that contributed most to the quality of life in the twentieth century. To announce the results, it invited Neil Armstrong, the engineer-turned-astronaut whose small step on the moon symbolized a giant leap for mankind. Armstrong was disappointed that space flight came in twelfth on the engineers' list of great achievements. However, based on the criterion of judgment, he had to agree: "While the impact of seeing our planet from afar had an overpowering effect on people around the Earth and provided the technology for tens of thousands of new products, other nominees were judged to have a greater impact on worldwide living standards."[59]

Armstrong explained how technological achievements pervade our daily life. Many are so familiar now that we take them for granted, forgetting that they were deemed miraculous only a hundred years ago. We are healthier than our ancestors were, thanks to improved diet, public hygiene, and health care, which owe much to agricultural mechanization, environmental works and pollution control, mass-produced medicine and advanced medical instruments, such as pacemakers and CAT scans. We flip a switch and get light, a hot stove, or power to drive electrical and electronic appliances. We turn a faucet to get clean water and push a lever to flush out waste for proper treatment. We are more mobile and connected. Cars, highways, and petroleum make travel accessible and enjoyable for most people. Airplanes shrink distances even more. With telephones and wireless, people thousands of miles apart talk as if face to face. Radios, televisions, and the Internet bridge cultural divides by disseminating news and information from all around the globe. Computers help us to organize the information and use it with unprecedented efficiency to solve difficult problems. Air conditioners keep us cool, and refrigerators keep food from spoiling on sweltering days. Technologies bring not merely convenience and luxury. By making the material conditions of life more affordable, they free an increasing portion of humankind from the bondage of poverty and enlarge the range of choices. Nor are the impacts of technology confined

Table 4.1   U.S. gross domestic product in 2000 by industry, in billions of dollars and as percentage of total GDP; employment in each industry; percentage of engineers, computer professionals, and natural scientists working in major sectors.

Industry	Gross product		Workers (millions)	% Workers in industry[a]		
	Bill. $	% GDP		Eng.	Comp.	Sci.
Agriculture	136	1.4	3.3			
Mining	127	1.3	0.6	1.2	0.3	3.3
Construction	464	4.7	6.8	2.4	0.3	0.1
Manufacturing	1,567	15.9	18.4	45.6	14.0	19.0
Durable goods	902	9.1	11.1			
Metal products	109	1.1	1.5			
Machinery	168	1.7	2.1			
Electrical equipment	181	1.8	1.8			
Transportation equipment	182	1.8	1.8			
Instruments	64	0.7	0.9			
Nondurable goods	665	6.7	7.3			
Food	137	1.4	1.7			
Chemicals	191	1.9	1.0			
Public utilities	825	8.4	7.1	5.3	4.4	1.1
Transportation	314	3.2	4.6			
Communications	281	2.8	1.7			
Electricity, gas	230	2.3	0.9			
Trade	1,568	15.9	30.5	3.2	6.6	1.2
Finance	1,936	19.6	7.6	0.8	12.2	0.2
Service	2,165	21.9	40.8	26.7	46.7	38.8
Business service	572	5.8	9.9			
Health service	547	5.5	10.2			
Legal service	133	1.4	1.0			
Education	789	8.0	2.6			
Engineering, management	306	3.1	3.5			
Government	1,216	12.3	20.6	11.4	8.1	31.9
Total	9,873	100.0	135.7			

Source: Census Bureau, Survey of Current Business, November 2001; *Statistical Abstract 2001*, table 783; Bureau of Labor Statistics, *Monthly Labor Review*, February 2002.
a. Numbers do not add up to 100 percent because some are self-employed.

to the material. More valuable than the spin-off products of the space program are the images of our home from space. The view of ourselves from a radically different perspective inspires us to ponder anew our position in the cosmos, to appreciate the fragility of spaceship Earth, and to treasure our shared life on it.

The Economic Arena for Technological Activities

Technology owes a great deal to science, but engineering does play a vital role without which scientific discoveries could not have the practical impact that they have. Contributions of various branches of engineering are discussed all through this book, but to gain an integral picture, let us glance at its major arena, society's productive activity. Table 4.1 breaks down the U.S. gross domestic product by industries.

The manufacturing sector has traditionally been the bellwether of the economy, but it is being challenged by the service sector. These two sectors, which sponsor almost 90 percent of the nation's industrial R&D, are home to almost three quarters of the nation's working engineers. The contributions of engineers to the construction, chemical, information technology, and engineering service industries are discussed elsewhere in the book. Here we look at those industries and branches of engineering that are inadequately covered. Although they attract less publicity, they display several hallmarks of engineering: stewardship of society's infrastructure; evolutionary technological improvements; and the distinction among product, process, and industrial engineering.

The Infrastructure for Energy

Energy is the lifeblood of the modern economy. Its production, conversion, and transportation constitute perhaps the most basic of society's infrastructures. Production includes exploration, extraction, and refining primary energy sources into usable fuels. Conversion turns fuels into electricity or mechanical work. Transportation brings energy through pipelines, electricity grids, and other vehicles to consumers. Each area calls for expertise in several branches of engineering and science. Crudely speaking, extraction of fossil fuels is chiefly the job of the mining industry, with its geologists and mining engineers. The petroleum-refining industry employs many chemists and chemical engineers. Design of energy conversion engines falls to mechanical engineers. Electricity is the domain of electrical engineering. Civil engineers build dams for hydroelectricity, and nuclear engineers build nuclear power plants.[60]

Fossil fuels—coal, oil, and natural gas—constitute more than 80 percent of the primary energy used in America. During the energy crisis of the 1970s, predictions abounded that the oil supply would peak by the end of the century and afterward fall behind demand. No one is so complacent as to think that oil reserves will last forever, but the day of oil shortages is continually being pushed back by technological progress, including higher efficiency in energy usage. Large and easily recoverable oil fields have mostly been exploited, but more reserves, in smaller fields, are being discovered with the help of advanced surveying technologies such as three-dimensional seismic imaging. Advanced extraction technologies make extracting oil from these fields cost effective. A directional drill can twist about while it drills underground, with sensors near the drill bit reporting what it hits. Engineers can monitor progress, steer the drill, change its direction from vertical to horizontal if necessary, and thus home in on pockets of oil inaccessible by vertical drilling. Powerful computers and geophysical models process seismic and sensor data in almost real time, enabling engineers to track the complex flow in a reservoir while it is being pumped and thus optimize operation. Methods such as steam or liquid carbon dioxide injection have also been developed to extract more oil from wells previously thought exhausted. These technologies not only increase the oil recovery rate but also reduce the impact of drilling on the environment. The next frontier for petroleum engineering lies in the deep, in opening up oil fields lying beneath more than 10,000 feet of water.[61]

Burning fossil fuels, biomass, and wastes inevitably produces greenhouse gases, which may cause global climate change. Stabilizing and reducing greenhouse gas emissions require a great variety of technologies.[62] Many people are enthusiastic about clean, renewable energy sources such as water, solar, wind, hydrogen, and geothermal power.[63] Technology has lowered the cost of efficient wind turbines and solar panels to competitive levels. Wind turbines with rotors as long as the wing span of a jumbo jet, originally developed for aeronautics, can generate megawatts of power. They are entering commercial service, but only in niche markets.[64] Hydroelectricity is still the only major clean, renewable energy source, although dams do have controversial environmental impacts. The remaining major energy source, nuclear power, emits no greenhouse gases but poses problems of radioactive waste disposal, and has been plagued by high capital costs and safety problems.

Revolutionary and Evolutionary Technologies

The fuel cell, which produces electricity by reacting fuel hydrogen with oxygen and discharging water as waste, is seen by many as the silver bullet that will provide unlimited, cheap, and clean energy. Probably it will in some distant future, but ideologues who invoke it to proclaim the imminent end of the gas-powered car have missed the value of incremental improvements.

The fuel cell was invented in 1839 by a Welsh judge and was used by the National Aeronautic and Space Agency (NASA) in the early 1960s for spacecraft. Its basic principles remain much the same today. The difficulty lies in making it commercially competitive. It would require numerous improvements not only in the fuel cell itself but also in many other automotive components, as the car would have to be totally redesigned. Even when the fuel cell–powered car is ready for the road, its popularization will call for hydrogen stations in numbers rivaling gas stations. For these and other reasons, a comprehensive report assessing automotive technologies up to the year 2020 is rather skeptical about the fuel cell, especially in view of the continuous fine tuning of the internal combustion engine.[65]

Revolutionary technologies such as the fuel cell make sensational stories for journalists and historians. Their merits need not be further emphasized. What deserves notice are the analyses of economists, who in studying steamships and numerous other cases have observed that although minor improvements on a technology have low visibility, their cumulative effect can rival or exceed the spectacular debut of that technology.[66] Just like compound interest, which over time can produce a fortune for a lender, persistent incremental improvements can work wonders. An annual increase in fuel efficiency of a few percent does not sound exciting, but over time it changes the picture significantly. On this point economists are at one with engineers, who excel not only in introducing revolutionary technologies but also in the evolutionary upgrading of mainstream technologies, often in response to consumer demand. Technological evolution has reduced the average fuel consumption per car by 50 percent in the past three decades. Although hardly noticed by journalists, such progress raises the bar for the fuel cell.

Improving a matured technology is anything but trivial, partly because it is already so good. A car or a ship is a complex system with many components, in which many inventions and innovations occur without fanfare. Because components of the car must work together, changes in one affect the performance of many others, which must also be changed accordingly to optimize the performance of the car as a whole. To fine tune a product and reduce the cost of its production, engineers make inventions that squeeze some gain here, adjustments that shave some waste there, sometimes redesign the whole to integrate numerous advances. Away from the limelight, technology evolves incessantly. Moore's law on integrated circuits, discussed in section 3.1, illustrates the potential of technological evolution.

Technological evolution partly explains why industries that are no longer on the cutting edge continue to invest in R&D. The automobile industry spends a smaller percentage of its revenue on R&D than such high-tech industries as aerospace and electronics, but it still spends significantly. In dollar amounts, Ford and GM topped all American companies in R&D expenditures in 1997 (see Appendix B for more details). Some of it goes to developing revolutionary technologies such as the fuel cell, which has already absorbed billions of dollars. Most goes to fine tuning mainstream cars and their manufacturing processes. You can appreciate the results by comparing the performance and equipment of the cars that you have owned over the years. Furthermore, as cost-effectiveness increases, prices drop, making technological goods affordable to a greater number of people.

Product, Process, and Industrial Engineering

An outstanding characteristic of modern manufacturing is the mass production of goods with considerable complexity, quality, variety, reliability, and affordability. Manufacturing caters to wide markets; involves many workers, machines, and supply sources; and may have extensive environmental impacts. Manufactured products roughly fall into two classes, as shown in Table 4.1. Durable goods, such as refrigerators and cars, are intended to last for years; nondurable goods, such as food and gasoline, are used up continuously. Both demand superior product designs and effective production processes, although the former tends to be more visible in durable goods and the latter in nondurable goods. Process engineering for nondurable goods manufacturing was considered in the context of chemical engineering in section 2.4. Here we focus on durable goods.

The two largest groups of engineers, mechanical and electrical, are most active in durable goods manufacturing. Computer engineers and systems analysts are important, but they also scatter in many sectors, including finance and service. Although one may expect to find more mechanical engineers in the machinery industries and electrical engineers in the electronics industries, the lines are not clear at all. Much electrical equipment has mechanical parts. Conversely, electrical and electronic systems consume about 20 percent of the total cost of an average 2000-model midsize passenger car, and their share is expected to rise to 30 percent by 2009. Up to 40 percent of the cost of a new aircraft goes to avionics, which consists of communication, navigation, radar, autopilot, and other systems. Most of these systems use some computer chips and embedded software, so that some people describe cars and airplanes today as computers on wheels or wings. In short, manufacturing demands the collaboration of engineers with diverse expertise.

Product designs—be the products Internet routers, wrist watches, medical instruments, robot arms, jet engines, communications satellites, or oil tankers—are paramount for manufacturing. Many products serve special functions, and designing them often requires specialized knowledge. To meet the challenge, there are specialty branches of engineering. Take, for example, two that use biology, along with other scientific knowledge. Combining physical and biological knowledge, *agricultural engineers* develop agricultural machines, soil conservation schemes, and agricultural product processing. *Biomedical engineers* design equipment for health care, including diagnostic instruments for ultrasound, electrocardiograms, computerized tomography, and other imaging technologies; therapeutic instruments, such as lasers for surgery and machines for kidney dialysis; artificial implants, organs, and prostheses. The National Academy of Engineering estimates that more than 30,000 engineers work in various areas of health technology. Their ranks are expected to swell as scientific advancement renders more biological phenomena susceptible to reliable manipulation. In agriculture and health care, two areas not usually associated with engineering, we see the prevalence of technology in all corners of modern life.[67]

"Design for manufacturing," a popular slogan among engineers, indicates the close relationship between product and process engineering. A fantastic product design is a failure if it cannot be manufactured for a reasonable cost. Conversely, manufacturing cannot run smoothly without the cooperation of product and process engineers. Both are mandatory, but their relative weight can shift according to market conditions. Consider, for example, two mottoes in the aerospace industry. "Faster, higher, farther," the chief aim of aeronautic engineering in the cold war era, has now become "faster, cheaper, better." The old maxim concerned only product performance; "faster" referred to the speed of flight, the time to target. In the new maxim, "faster" refers to the speed of design and production, the time to market. Of course product performance cannot be neglected, as the "better" in the new motto reminds us. Fast and cheap production becomes a waste if it fails to deliver products with satisfactory performance. Thus NASA, which tried to build two Mars probes for the price of one, got nothing but criticism when its Mars climate orbiter went off course and the Mars polar lander crashed.[68]

Manufacturing depends on three major factors: labor, capital, and material. Gasoline production requires a lot of capital to build petroleum refineries and much engineering to design the refineries for maximum use of expensive crude oil, but an operating refinery requires relatively few workers to churn out the gasoline. Car production, too, requires much capital and engineering design, but it also requires many workers in parts production and assembly. Although the transportation equipment industry is slightly smaller than the chemical industry in terms of product, it employs 80 percent more workers. Such labor-intensive manufacturing accentuates the importance of organizational technology in the production process. This is the task of *industrial engineering*, which plans effective deployment of production workers; coordinates supply chains for parts and materials; designs factory layouts for effective interfaces among people, machines, and materials. Germinated in the nineteenth century, it grew and changed tremendously after World War II with the infusion of operations research and other management techniques.

One important area for industrial engineering is the transportation equipment industry, about 60 percent of which is devoted to automobiles and parts; 30 percent to aircraft, missiles, and space vehicles; and the remaining 10 percent to ships and railroad equipment. Cars and aircraft are complicated machines with many parts. Their manufacturing involves numerous product parts and production processes that must be properly organized. Detroit's three big automakers depend on an army of some 12,000 parts suppliers grouped into several tiers. Large suppliers in the top tier manufacture automotive systems such as the drive train, and they, in turn, depend on lower-tier suppliers for components such as steering wheels and alternators. A similarly complicated supply chain exists in the aerospace industry. Designing efficient supply chains for maximum productivity requires coordinating efforts not only within a corporation but also among many corporations. These tasks are taken up by the organizational technology in industrial engineering.

5

INNOVATION BY DESIGN

5.1 Inventive Thinking in Negative Feedback

"It was a mental state of happiness about as complete as I have ever known in life," recalled Nikola Tesla about the days in 1882 when he invented the polyphase system for generating, transmitting, and using alternate-current electricity.[1]

"Wilbur and I could hardly wait for morning to come, to get at something that interested us. *That's* happiness," wrote Orville Wright about the last weeks of 1901, when the brothers performed wind-tunnel experiments that unlocked the secrets of lift. They retrospectively rated the experience the peak of their intellectual excitement, unmatched by subsequent success and fame from building the first manned and powered aircraft.[2]

Countless reports support the observation of electrical engineer Jerome Wiesner: "Technical and scientific work is usually fun. In fact, creative technical work provides much the same satisfaction that is obtained from painting, writing, and composing or performing music."[3] Engineers and scientists are more absorbed by their jobs than many people, because they love what they do. One survey found that engineering students were more focused on their subject than students majoring in other areas, less because their subject was difficult than because they found it fascinating. The commonplace of engineers happily working long hours in high-tech firms is captured by the title of a book on the development of the Internet: *Where Wizards Stay Up Late*. It is nothing new; the line was culled from an old poem describing the working ethos at Los Alamos. A similar fire burned in the engineers who routinely worked eighty-hour weeks to develop the Apollo Lunar Module, a feeling recalled by participants as "a sense of mission" or "spirit in that engineering group."[4] Enthusiasm may be amplified by group dynamics, but its root is spontaneous and occurs as frequently in lone workers. Tesla worked daily from 10:30 in the morning to 5:00 the following morning in the laboratory of Edison, who was himself a workaholic. Henry Ford described how he designed internal combustion engines in his spare time while working as an engineer at Detroit Edison Electric: "Every night and all of every Saturday night I worked on the new motor. I cannot say that it was hard work. No work with interest is ever hard."[5]

Joy in the work itself, before rewards of social acclaim and financial return, is the "existential pleasure of engineering" as vividly described by Samuel Florman in recounting his own personal experience.[6] The feeling is shared by scientists. Shannon cited pleasure as one of three chief motivations of engineering research, besides intellectual curiosity and the urge to improve things.[7] Similarly, Feynman cited intellectual enjoyment as one of the three major values of science, besides the ability to do things and the experience of uncertainty.[8] Perhaps more than anything else, intellectual excitement attracts engineers and scientists.

Enthusiasm does not imply wantonness, and technical creativity is not whimsical. Science and engineering are both objective and concerned about reality. Engineers have some leeway in design decisions to express their idiosyncrasies, and many great bridges bear the personal stamp of their structural engineer. The subjectivity, however, is severely constrained by realistic considerations, both physical and socioeconomic.

Do constraint and discipline stifle creativity? Not necessarily, if creativity is distinguished from capricious flights of fancy. Creativity in engineering and science is not inferior because of discipline, for discipline is a part of any worthy creative activity. A random splash of color or concatenation of words would probably yield a new image or sentence, but without originality. Leonardo da Vinci explained why rules are nec-

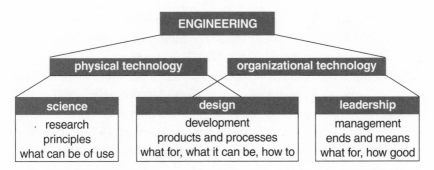

Figure 5.1 Engineering activities and concerns.

essary for good painters: "These rules will enable you to have a free and sound judgment; since good judgment is born of clear understanding, and a clear understanding comes of reasons derived from sound rules, and sound rules are the issue of sound experience—the common mother of all the sciences and arts."[9] Poets know this equally well. William Wordsworth observed: "For all good poetry is the spontaneous overflow of powerful feelings: and though this be true, poems to which any value can be attached, were never produced on any variety of subjects but by a man, who being possessed of more than usual organic sensibility, had also thought long and deeply."[10] Johann Wolfgang Goethe put it succinctly: "None proves a master but by restraint."[11]

Restrained Creativity of Engineering

Engineers create artifacts to serve people. Besides having ingenuity, they need to think long and deep in two areas. To design complicated things requires knowledge about the physical world; to serve people properly requires knowledge about the social world. The dual physical and human dimensions are expressed in three engineering activities: science, design, and leadership, as illustrated in Fig. 5.1. They enlarge three repositories of technology as a productive capacity: knowledge, implements, and human resources.

Natural scientists explore what is. Engineering scientists explore what can be—or, more exactly, what can be of practical use to human beings as limited by natural laws and other factors. On the level of general principles, the crucial distinction between engineering and natural science is not the hypothetical implication of *what can be* but the qualification *of use,* although it is often abbreviated. The *what is* of natural science asks for not catalogs of existing things but physical laws, which cover possible as well as actual phenomena and support contrary-to-fact hypotheses. Its peculiarity as the science of natural phenomena lies in its exclusion of human subjectivity from its objective content so that it explains nature as it is. In contrast, engineering science aims at actual and potential artificial systems that render practical service. Therefore it explicitly incorporates purposive concepts as indispensable elements in its objective content.

Purposive concepts in engineering fall roughly into three groups: functional concepts asserting what the artifacts are for, procedural concepts asserting how to bring the artifacts into being, and evaluative concepts assessing how well the artifacts perform to satisfy their intended purpose. The questions *what for? how to?* and *how good?* are present in all aspects of engineering. Stated abstractly in engineering science, they become more concrete in design and leadership.

As typical of science, *what is* and *what can be* refer less to particular systems than to broad types of systems covered by general principles that chart out huge areas. For instance, information theory explores the bounds of what can be achieved for reliable telecommunication in general. Engineering design narrows the question of what can be to what *it* can be. *It* can refer to a singular system, such as a cell phone network for a particular region, or to a particular kind of artifact, such as a high-speed modem for connecting computers to telephone lines. *It* signifies particularity, which is central to design. While the scientific answer to *what it is* provides factual descriptions, the design answer to *what it can be* reveals the creative freedom peculiar to engineering. Of course the system under design must satisfy many requirements, but even stringent requirements usually leave open a range of options, the choice among which rests on the discretion of design engineers. In examining options, assessing credits, fixing details, and proceeding from *what it can be* to *what it will be,* design engineering is as much a technical as a decision-making activity.

In designing a technological system, engineering is creative under many constraints. Often the constraints are not clearly articulated but must be ferreted out by the engineers themselves to get a clear conception of the product. First of all, the product is ordered for certain purposes, which are expected to be served by the product's performing certain functions. To ascertain the purposes and requisite functions is to get substantive answers to the question *what for?* Then performance criteria are formulated so that candidate designs can be evaluated as to *how good* they are—how well they satisfy the requirements. A design that performs well serves little purpose if it cannot be manufactured within cost and other constraints, so production processes—or answers to the question *how to?*—come into consideration. Design constraints also come from questions such as where to build a system and, more frequently, when it must be delivered. Often questions arise as to who pays for it and who benefits from it, for instance whether to make it easier to manufacture or to maintain, which will determine the relative costs for the manufacturer and the operator. In sum, design engineers have their eye on the five p's: the intended product, its purpose, performance, production process, and the people whom it is intended to serve or work with, for they are the ones who will decide its ultimate fate. All these aspects introduce constraints, which are often inconsistent. Engineers must make complex trade-offs and choices, often in the face of uncertainty. To understand how they make practical decisions in designing technological systems is crucial to understanding technology.

Answers to the questions *what for?* and *how good?* ultimately depend on the choices of society at large. Because products extend their services into the future, many elements of which cannot be foreseen, engineered systems involve certain risks and uncertainties, and stakes rise as technological complexity and potency soar. Engineers can make things extremely reliable, but not absolutely so; perfection is beyond their human capability. Furthermore, higher reliability usually comes with higher cost, which society must decide whether it is willing to bear. Engineers often ask: how safe is safe? how good is good enough? Considerations such as economic viability, social acceptability, political regulation, and environmental benignity involve many groups of people. Their wants are sometimes technically unrealistic and their claims incompatible. Engineers must critically balance conflicting claims, assess risks, design plans, and explain them to the public so that the public can make difficult choices. In these tasks, their leadership becomes prominent.

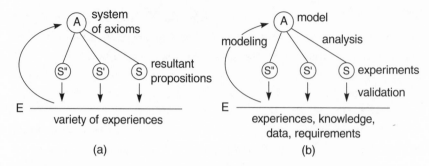

Figure 5.2 (a) A diagram in Einstein's letter to Solovine illustrating his epistemological view of scientific thinking. (b) Similar types of thinking in engineering.

Types of Thinking in Science and Engineering

Besides physics, Albert Einstein also thought long and hard about the epistemology of science. He wrote a lot about it. Once, in a letter, he drew a little diagram as in Fig. 5.2a. With it he discussed the relation between scientific theories and reality. Scientists are familiar with observations and experiences, represented by E in the figure. He wrote, "There exists no logical path leading from E to A," what he labeled as "axioms" but would also include theories, models, concepts, and hypotheses. A scientific theory is usually a system of general statements that cover a wide range of cases by abstracting from the disparate details of individual cases. We can think in general terms, but as physical beings we encounter the world only in particular situations. Therefore, to test a general hypothesis, we must derive from it propositions applicable to this or that particular case, marked S, S', or S'' in the diagram. Einstein explained: "From the A are deduced, *by a logical path*, particular assertions S that can claim to be exact. The S are brought into relation with the E (testing by experience). This procedure belongs also to the extra-logical (intuitive) sphere."[12]

Regarded as a loop, Fig. 5.2a looks like the hypothetical-deductive model expounded in the philosophy of science. However, Einstein did not write as if he was talking about four stages of a process. Rather, he explained the epistemological characteristics of four *types* of thinking. For convenience of reference, I call them experience (E), creation

$(E{\rightarrow}A)$, deduction $(A{\rightarrow}S)$, and experimentation $(S{\rightarrow}E)$. Of the four, only deduction can be—not always is—exact, because the path from a general axiom to a particular proposition connects concepts only. In thinking that connects concepts to experiences of reality, exactitude and absolute certainty are illusory.

Einstein concluded his letter: "The quintessence is the always problematic connection between the world of ideas and that which can be experienced." The epistemological uncertainty is acknowledged by many scientists and engineers. Electrical engineer David Slepian expounded it in tackling a paradox in the bandwidth theorem.[13] Feynman regarded coping with uncertainty a major value of science: "The scientist has a lot of experience with ignorance and doubt and uncertainty, and this experience is of very great importance, I think. When a scientist doesn't know the answer to a problem, he is ignorant. When he has a hunch as to what the result is, he is uncertain. And when he is pretty darn sure of what the result is going to be, he is still in some doubt. We have found it of paramount importance that in order to progress we must recognize our ignorance and leave room for doubt. Scientific knowledge is a body of statements of varying degrees of certainty— some most unsure, some nearly sure, but none *absolutely* certain."[14] Experiences with risk and uncertainty sink even deeper in engineers, who are morally and legally liable for the safety of their products.

Einstein discussed not content of natural science but the characteristics of scientific thinking. The same types of thinking can take on other content, for instance engineering theory and design. This is illustrated in Fig. 5.2b, in terms familiar in the engineering literature on modeling and simulation.

Theories, represented by A, exist on many levels of generality. On the most general level are sweeping theories such as quantum mechanics in physics or information theory in engineering, which establish arching but abstract conceptual frameworks within which the bulk of scientific and engineering activities proceed. Less general theories specify some substances that narrow the scope of investigation, for example the quantum theory of atomic structures or coding theory for reliable transmission of information on noisy channels. Focusing further, we have the theory for the helium atom or the turbo code. Such theories are often called *models,* partly because of their narrow focus on a small piece of reality. General theories are rare in any field, but models abound.

Most models in engineering are not covered by sweeping theories. They are conceptual representations introduced to study small types of systems, such as highway or Internet traffic. Modeling also involves the four types of scientific thinking, albeit with a smaller scope. Engineers, with their knowledge and experience, first examine a situation, which often contains requirements stipulated by the intended users of the systems. To create a model that captures important features of the situation, they introduce concepts to represent and quantify not only physical factors but also intangible ones, such as reliability. As in science, engineering modeling is a creative act; it may have heuristics but no definite logic. When a model A is formulated, often in mathematical terms, it is analyzed and its parameters varied to yield specific instances S and S'. In simulation models, this step is often called experiment design. Then the results obtained for those instances are checked against the real situation to decide whether the model is acceptable, a process engineers call model validation. Valid models play a crucial role in the engineering design of systems that will be implemented physically.

Creativity and Insight in Science and Engineering

The leap from experience E to model A in Fig. 5.2 occurs in many facets of engineering and natural science. Einstein explained that the leap is intuitive; logical methods such as induction may help but are never sufficient. This point he stressed in many writings, as in examples of atomic theories: "The theoretical idea (atomism in this case) does not arise apart from and independent of experience; nor can it be derived from experience by a purely logical procedure. It is produced by a creative act."[15]

Engineers must conceive a system first before building it. Similarly, Einstein remarked: "It seems that the human mind has first to construct forms independently before we can find them in things."[16] Most patterns and regularities are not obvious in natural phenomena that are bewilderingly complex. The sun is obvious, but the source of its energy is not. Physicists have to hypothesize nuclear fusion before verifying it with measurements, just as nuclear engineers conceive of fission reactors before realizing the designs physically. Even in mathematics, the paradigm of deductive disciplines, conjecture is indispensable. Mathematician George Pólya explained: "Finished mathematics presented in a finished form appears as purely demonstrative, consisting of proofs only. Yet mathematics in the making resembles any other human knowledge in the making. You have to guess a mathematical theorem before you prove it; you have to guess the idea of the proof before you carry through the details."[17] Given a set of axioms, a simple computer program can generate endless theorems that are almost all junk. Mathematicians have to discern the few significant theorems, and that calls for intuition and insight. Similarly, successful science depends on discerning important phenomena worthy of investigation; successful engineering, on discerning devices worthy of innovation. That insightful discernment is the creative element crucial to science and engineering.

Much has been made of the prowess of mathematics and computation in research and development. Important as they are, they do not and cannot stand alone in engineering and science, as structural engineer Henry Petroski wrote: "Mathematics and science help us to analyze existing ideas and their embodiment in 'things,' but these analytical tools do not in themselves give us those ideas."[18] Galileo distinguished the astronomer "merely as a calculator" from the "astronomer as a scientist." The calculator grinds out planetary orbits under given hypotheses. The scientist is dissatisfied with merely saving the appearances with arbitrary hypotheses and demands an explanatory system wherein "the whole then corresponded to its parts with wonderful simplicity."[19] His aesthetic evaluation accentuates insight and other abilities embodied in engineers and scientists, which can be partially articulated but not totally. These are explained by chemist Michael Polanyi: "I regard knowing as an active comprehension of the things known, an action that requires skill. Skillful knowing and doing is performed by subordinating a set of particulars, as clues or tools, to the shaping of a skillful achievement, whether practical or theoretical."[20] Skill and intuition, tacit and inalienable from human activity, explain why the human resource is the cornerstone of technology that robots cannot supplant.

Skills can be improved by effective heuristics. Pólya described many heuristic principles and illustrated them with examples in mathematics:

abstract and simplify; guess and test; divide and conquer; isolate and combine; generalize and specialize; reorganize and regroup; work backward and forward; narrow and widen the scope; strengthen and weaken the conditions; if you cannot solve the proposed problem, look around for an appropriate related problem.[21]

Systems engineers Mark Maier and Eberhardt Rechtin collected hundreds of heuristics from aerospace, electronics, and software industries. They found that four recurrent themes stood out:

1. Don't assume the original statement of the problem to be necessarily the best or even the right one.
2. In modularization, choose elements with low external complexity and high internal complexity, so that the elements are as independent as possible.
3. Simplify, simplify, simplify.
4. Build in and maintain options as long as possible in the design and implementation of complete systems. You will need them.[22]

Most of these engineering heuristics are also effective in natural science, because scientists, too, struggle with complex phenomena. The second heuristic, for example, is apparent in independent-particle approximation and other models. There, physicists reformulate a system with tightly coupled elements by introducing new loosely coupled entities that absorb complexities previously regarded as external relations into their internal structure.[23]

What to Discover, What to Invent?

Luck is always a good ally but, unlike in casinos, it does not dictate in science and engineering. Research is not chance observation, and invention is not accidental ideation. When your mind is not prepared, you would not recognize a regularity or design even if you stumbled on it. It is significant that the most popular heuristic collected by Maier and Rechtin concerns how to frame problems. The image of engineers and scientists as good problem solvers obscures the fact that researchers are not students for whom professors have designed problem sets. They must themselves identify what needs to be and can be solved, conceptualize relevant factors, and spell them out in formulating a problem. Clear problems direct effort and indicate certain criteria for acceptable solutions. To make a flying machine is more a vision than a problem; it is too vague. To build a machine that flies like a bird is a trap from which would-be inventors freed themselves only after many failures. Problems are not fixed but often evolve as knowledge increases. Even in a field where most researchers have a good hunch where the important problems lie, delineating a sharp and immediately tractable problem is not easy.

Venturing into the unknown, researchers initially rely heavily on groping around. Edison recounted how he worked: "I would construct a theory and work on its lines until I found it untenable, then it would be discarded and another theory evolved. This was the only possible way for me to work out the problem."[24] Searching through possibilities, guessing at solutions, and narrowing down the focus are tasks that belong to rational empiricism. It is different from blind cut-and-try, as Feynman observed: "It is not unscientific to make a guess, although many people who are not in science think it is."[25] Scientists and engineers gain knowledge when a guess is proved wrong. By figuring out why something fails, they rule out not only the tried case but also a class of possibilities, get a better feel for the difficulties and the requirements of what might work, choose a better angle of approach, and perhaps reframe the problem. Those experienced in a field have tried so many cases and examined the field from so many angles that even a complicated problem becomes quite transparent and they can spot at a glance faulty proposals and refuse to waste their energy on them. Edison's remark that "ninety-nine percent of genius was knowing the things that would not work"—a knowledge gained through perspiration—resonates with scientists. Werner Heisenberg, known for his uncertainty principle, said: "The research worker who is thoroughly familiar with the field in question is probably of the just conviction that he could at once refute any false theory. If the new proposal appears to be a genuine possibility in which the earlier difficulties are avoided, if it does not immediately run into insoluble problems, then it has to be the right proposal."[26]

Framing the Right Problem and the Right Strategy

Finding the right problem and the right approach were decisive in Harold Black's invention of negative feedback. When Black joined the Bell

System in 1921, AT&T was striving to expand the capacity of long-distance telephony. Its first transcontinental cable, opened in 1914, carried a single voice channel per pair of open wires and overcame transmission loss with three to six vacuum-tube amplifiers spaced over 3,000 miles. Putting more channels on the wires would increase loss and hence require more amplifiers along the cable. Available amplifiers distorted a signal, so that a large number of amplifiers in tandem would distort a signal beyond recognition. The lack of good amplifiers was a major roadblock attacked by many Bell engineers, including Black. Their efforts had little success.[27]

In 1923, Black attended a talk by Steinmetz and was impressed by "how Steinmetz got down to the fundamentals." Fundamentals reveal general grounds and broad considerations, which may suggest new possibilities invisible to those trapped in specifics. Inspired, Black broadened his view on the project, critiqued his previous strategy, and revised his problem. He realized that the fundamental problem was to obtain low-distortion transmission over long distance. A low-distortion amplifier was a *means* to this end and should not be mistaken as a basic *aim* in itself. Stonewalled, it was time to seek alternative means. Thus he reframed his problem. Instead of trying to *prevent distortion* in an amplifier, he accepted distortion in amplification as a necessary evil and tried to *cancel the distortion after amplification*. With this new strategy, he invented the "feedforward" amplifier, based on the idea that distortion can be measured, and the measured value can then be subtracted from the distorted amplifier output to obtain the desired distortion-free value. To measure the amplifier distortion, Black used a second amplifier, reduced its output to the input level, and subtracted away the input. The system worked and proved that low distortion can be achieved by cancellation rather than by prevention.

Feedforward control is used today to compensate for distortions that can be measured by a sensor. However, it is not practical for signal amplification; the gains of the two amplifiers must be so carefully balanced that they must be tuned every hour, around the clock. Black recalled how he struggled to improve his invention: "Nothing came of my effort, however, because every circuit I devised turned out to be far too complex to be practical. I was seeking simplicity."[28]

Black's identification of the problem and his goal of simplicity paid off. A spark leapt on a ferry ride to work one morning in 1927, but how, he could not say: "All I know is that after several years of hard work on the problem, I suddenly realized that if I fed the amplifier output back to the input, in reverse phase, and kept the device from oscillating, . . . I would have exactly what I wanted: a means of canceling out the distortion in the output." In a negative feedback amplifier, if the loop gain is much greater than one, then the overall gain does not depend significantly on the behavior of the distorting amplifier. Furthermore, the feedback circuit can often be built from robust passive elements, such as resistors and capacitors, in contrast to the finicky amplifier in the feedforward circuit.

As usual, developing the invention into a useful product required ten times the effort of invention. Black was joined by Harry Nyquist, Hendrik Bode, and other Bell engineers to solve stability problems and other difficulties. Partly because of their efforts, a transcontinental cable carrying 480 telephone channels using 600 repeaters in series was opened in 1941. Since then, negative feedback has found wide employment, not only in communications and artificial control systems of all kinds, but also in illuminating natural phenomena.

Simplicity in Combating Complexity

Black's road to negative feedback had at least two major landmarks. The first, changing the problem from preventing to canceling distortion, exemplifies Maier and Rechtin's first heuristic: maintain a broad perspective. Shannon explained the heuristic in his reflection on creative thinking: "Change the viewpoint. Look at it from every possible angle. After you've done that, you can try to look at it from several angles at the same time and perhaps you can get an insight into the real basic issues of the problem, so that you can correlate the important factors and come out with the solutions."[29]

The second breakthrough to negative feedback resulted from Black's insistence on simplicity. Simplify, simplify, simplify, so clearly asserted in Maier and Rechtin's third heuristic, is a common theme in engineering design and scientific research. Since the fourteenth century, the law of conceptual parsimony known as Ockham's razor has been used to cut through the thicket of words masking dogmas and superstitions. Isaac Newton made it the first rule of reasoning in natural philosophy: "We are to admit no more cause of natural things than such as are both true and sufficient to explain their appearances. To this purpose the philosophers say that Nature does nothing in vain, and more is in vain when less will serve; for Nature is pleased with simplicity and affects not the pomp of superfluous causes."[30]

Physicists' talk about simplicity is sometimes mistaken for metaphysical assertions about nature. A closer look at their remarks reveals that simplicity is mainly a rule of *reasoning,* an epistemological guiding principle for concepts and theories by which we understand and represent natural phenomena, especially complex ones. As well as engineers, physicists know that causes are pregnant with effects of all kinds. A careless conjecture of causes to fill holes in a theory can have disastrous unintended side effects. Simplicity demands that theories go as far as possible in expunging both holes and danglers, thus explaining the greatest range of phenomena with the smallest number of causes. In this vein Einstein wrote: "It is the grand object of all theory to make these irreducible elements as simple as possible, without having to renounce the adequate representation of any empirical content whatever."[31]

Engineers strive to serve maximal functions with minimal forms. Simplicity is among the top engineering imperatives, variously known as the *principle of economy* or the *bane of overdesign.* Perhaps the most often quoted dictum is, "Keep everything as simple as possible, but not simpler," attributed to Einstein but never confirmed. Examples from practical experiences abound. "My effort is in the direction of simplicity . . . Start with an article that suits and then study to find some way of eliminating the entirely useless parts," Ford said in describing how he designed the Model T, which despite fierce competition was voted by automobile enthusiasts as car of the century.[32] "Keep it simple" was the direction Joseph Shea set for the Apollo Spacecraft Program in Houston: "Figure out how to do things as simply as possible, with as few ways as possible to go wrong." Thus, instead of creating a sophisticated material for the command module that would withstand both the extreme heat of reentry and the extreme cold of outerspace, engineers made the spacecraft rotate on its axis once a hour all the way to the moon and back. This simple "barbecue" mode saved the heat shield from extreme cold by toasting it regularly in the sun.[33]

The complex phenomena faced by engineers and scientists are often intractable without some simplification. The crux lies in what simplifications are made and how they are justified. Contrary to simplistic doctrines that neglect important factors, scientific simplifications capture most if not all essential factors in a wide class of complex phenomena. Shannon again explained: "A very powerful approach to this is to attempt to eliminate everything from the problem except the essentials; that is, cut it down to size. Almost every problem that you come across is befuddled with all kinds of extraneous data of one sort or another; and if you can bring this problem down into the main issues, you can see more clearly what you're trying to do and perhaps find a solution."[34]

Those who deck their simplistic ideas with bells and whistles brag about complexity; those who tackle real-world complexities strive for simplicity. Simplicity dispenses with frills and demands clarity, coherence, precision, and austerity. It is characteristic of scientific theories that represent complicated phenomena in mathematical forms so coherent they appear obvious to those willing to make the effort for understanding. It is also characteristic of engineering designs that serve complex functions so smoothly that even the most onerous tasks appear effortless. None of the theories or designs is trivial or easy. Understanding general relativity requires a graduate-level education, but the theory is lauded for its simplicity because, once mastered, its few general principles elucidate properties of the gravitational field otherwise hopelessly opaque. It is in grasping the most complex phenomena and tackling the most difficult problems that the simplicity of scientific concepts and the elegance of engineering solutions become most apparent. Simplicity resides in mastery of complexity.

Simplicity often goes together with beauty. Simplicity that lets the grandeur of physical reality shine through gracefully in the service of human welfare is the aspiration of all and the achievement of many engineering designs. Simple aerodynamic lines and other engineered functionalities integrated into the style of a car have endowed some models with enduring grace. In contrast, when American cars deviated from the engineering tradition and relied solely on artsy designers in the 1950s, the result was tacky tail fins. Aesthetics is most apparent in civil and structural engineering. The soaring arches, intricate trusses, and graceful suspension cables of bridges, although designed for utility, exude beauty. Architects may paint a bridge or sculpt its towers, but these refinements are minor. The Golden Gate Bridge would be sublime whether painted red or gray. In their unadorned structures, great bridges harmonize with land and water, fly across wide spans, carry heavy traffic, withstand gales and earthquakes, all with the appearance of perfect ease. Civil engineer David Billington rightly maintains that skyscrapers, great bridges, and grand roof vaults constitute structural art, a new art form derived from engineering, with efficiency, economy, and elegance as its basic principles.[35]

Naturalness is another characteristic of simplicity. A good scientific theory appears so coherent and free from contrivance it carries an air of inevitability. Newton's laws of motion seem now to be self-evident, but if they were, they would not have been so shocking in the seventeenth century. Simplicity testifies to the creative genius of scientists and engineers. Judge Townsend expressed it best in his ruling in favor of Tesla's patent application for polyphase alternating-current systems: "The apparent simplicity of a new device often leads an inexperienced person to think that it would have occurred to anyone familiar with the subject, but the decisive answer is that with dozens and perhaps hundreds of others laboring in the same field, it had never occurred to anyone before."[36]

5.2 Design Processes in Systems Engineering

When Socrates was asked about acting justly for justice's sake and not mere expediency, he said it would be easier to explain not on the small canvas of personal actions but on the large canvas of public policies.[37] Personal motivations are often unstated and messy, but people must marshal explicit and clear reasons when they try to persuade fellow citizens. Similarly, the thinking processes of individual engineers and scientists are mostly implicit and intuitive, and are seldom reported in research papers meant to convey results. In teaching, professors try hardest to articulate technical thinking. Articulation also occurs when professionals must coordinate their actions in large projects, a situation not unlike public discourse on appropriate policies. Large projects with a wide social impact that demand collaboration and negotiation among many parties are increasingly common in engineering. To handle them, a discipline known as *systems engineering* has emerged that aims to investigate and improve on the process of engineering design and development. It provides a large arena in which the rationales for engineering actions are revealed more clearly.

"Systems engineering is the design of a design," according to engineers at Grumman (now Northrop Grumman) who practiced it admirably in designing the Apollo Lunar Module.[38] Restraining the impulse to run off and do things right away, systems engineers step back to think about how best to proceed in creating complex systems. Like a method of research, design of designs tries to bring into explicit awareness fruitful and adaptable guidelines for technical thinking.

"Systems engineering is just plain common sense in that each concept, each step, is the reasonable thing to do. The value of the systems approach is that it allows you to bring all these common-sense ideas together in concert to focus on the resolution of complex problems in complex environments," said systems engineer Ralph Miles in his introduction to a series of lectures on the topic.[39] Common sense here does not refer to folk beliefs, such as that the sun revolves around the earth, many of which are refuted by scientific inquiry. Rather, it refers to human rationality, the ability that jurors exercise in deciding verdicts and that everyone exercises in attempting to weigh evidence and make judgments in daily decisions. Engineers and scientists consider problems that are more complicated and often far removed from daily experience. They proceed more systematically, wielding more powerful techniques to tease out evidence and deduce conclusions. Despite all their mathematics and instruments, however, they still rely on their human rationality, honed to be more critical and discerning but still fallible, as Einstein wrote: "The scientific way of forming concepts differs from that which we use in our daily life, not basically, but merely in the more precise definition of concepts and conclusions, more painstaking and systematic choice of experimental material, and greater logical economy."[40] Systems engineering helps engineers manage complex technological projects. As critically articulated and refined common sense, its rationale is also illuminating for personal thinking and management of individual life.

Concepts of System

The system is among the most common concepts in engineering. From the Greek word meaning organized whole, government, or a body of men or animals, *system* generally means a unity comprising mutually related parts, be they concrete or abstract, static or dynamic, natural or artificial, things or people. Systems are neither trivial nor chaotic. The first qualification excludes aggregates that are the mere sum of their parts, which are trivial for lacking the main source of complexity, relations among the parts. The second qualification rules out Thomas Hobbes's "state of nature," in which all are at war with all. In one of the earliest usages in English, Hobbes wrote in 1651: "By Systemes; I understand any number of men joyned in one Interest, or one Business."[41] An assemblage qualifies as a system only if the relations among its components exhibit at least some nontrivial order and cohesion susceptible to rational articulation.

Economist Adam Smith, a friend of James Watt, realized that "a machine is a little system." Railways and electrical power and telephone networks brought new significance to systems; a network is a system where the interrelations are more prominent than the attributes of the entities related. Network analysis became a major engineering task in the late nineteenth and early twentieth century.

The scientific analysis and development of systems was accelerated during World War II through two activities, operations planning and weapons development. Military operations require coordinating the movements and actions of large quantities of people and materials with speed and efficiency. Realizing that they could be represented and analyzed mathematically, the British initiated a method of "operational research" that mobilized "scientists at the operational level." After the war, it was broadened into systems analysis and pursued by several professions, including industrial engineering and management.[42]

Development of weapons systems is more in the domain of engineering. An antiaircraft defense system required the integrated functions of guns, radar, telescopes, searchlights, communication links, information processing machines, and their human operators. This project and other systems called for the cooperation of engineers and scientists from many areas. Control engineer and historian Stuart Bennett wrote that, among the influences of the war effort on engineering, "the primary one was the recognition of the need for, and adoption of, a systems approach to problems. This was partly through a recognition of the underlying commonality between apparently diverse pieces of equipment, and partly through a recognition that in order to achieve an overall goal it was necessary to consider not just the performance of individual units of a piece of equipment designed to achieve the goal but also how the units interact with each other."[43]

The war effort spawned three prominent meanings of system in engineering: systems theory, systems approach, and systems engineering. The names are used without much consistency, but the meanings behind them are clear. They emphasize, respectively, the science, design, and leadership aspects of engineering.

Recognizing the commonality beneath diverse types of equipment leads to *systems theoretic thinking,* which abstracts from *material* characteristics and concentrates on *functional* characteristics. In the theory of signals and systems, for instance, a system is simply something that performs a function in signal processing by transforming signals in a certain way. It does not matter whether the signal is acoustic or electrical and whether the transformation is mechanical or electronic. The theoretical results can be implemented either way, bringing into relief the functional commonality of disparate devices. Abstraction, concept formation, and mathematical representation are crucial scientific methods. Here they are applied in strategic engineering research for general types of useful functions. Systems theories such as optimal control are widely used in many branches of engineering to design systems of all kinds. Engineering also shares with operations research and management science a formidable mathematical toolbox containing linear, nonlinear, and dynamic programming; queuing, network, and game theories; simulation and other powerful computational methods. Although indispensable in engineering design, they can be examined on their own as scientific theories, which we will do in section 6.2.

While systems theoretic thinking operates on a high level of generality and abstraction, the *systems approach* attends to all details of a particular system—not details for details' sake but details of parts that must work together for the integral function of the system as a whole. The emphasis on distinctive parts in addition to the whole is crucial, although it is neglected in misrepresentations of the systems approach as mere holism that shuns parts analysis. To design a complex system, engineers modularize it into subsystems with proper interfaces, and then further analyze each subsystem into smaller components, until they get components simple enough to be designed to the last detail. Then they synthesize the designed components and subsystems into a single functioning system. Thus they master the system on many scales, plus their interconnections. In designing a chemical processing plant, for instance, chemical engineers consider everything from microscopic molecular dynamics through mesoscopic transport phenomena to macroscopic unit operations. In this multilevel view, the systems approach allows engineers to see both the forest and its trees clearly, as discussed in section 5.4.

Deep and thorough, the systems approach focuses on designing a system's internal structure, whose functions serve a certain objective. *Systems engineering* encompasses the systems approach within a broad and long-term view that includes two other factors: analysis and choice of objective, and analysis of the system's life cycle from cradle to grave. To tackle them, systems engineering integrates organizational and physical technologies.

Rise of Systems Engineering

Bell Telephone Laboratories' 1945 study of antiaircraft guided missile systems was regarded by members of its technical staff as "a milestone in systems engineering precisely because it was comprehensive enough to address a *whole* system—propulsion, aerodynamic control, steering strategy, warhead design, sensors and computers—to find an economical and effective design." This systems approach was not all, though. Members of Bell Labs continued: "In all of its military projects, Bell Labs has worked with customers to define their needs as completely as possible and to arrive at agreement on objectives . . . Gradually, the process of defining objectives and analyzing and evaluating alternative plans for their accomplishment has become an activity separately defined as systems engineering."[44] Far from being merely a set of tools with which to find means to given ends, as it is often caricatured, systems engineering was born with *analysis of ends* and *definition of objectives* as integral parts. This emphasis persists. Its civilian version was enunciated in 1949 by mechanical engineers Gordon Brown and Duncan Campbell: "We have in mind more a philosophic evaluation of systems which might lead to the improvement of product quality, to better co-ordination of plant operation, to a clarification of the economics related to new plant design, and to the safe operation of plants in our composite social-industrial community."[45]

G. W. Gilman, Bell Labs's director of systems engineering, offered a course on the subject at MIT in 1950. The University of Arizona established the first department of systems engineering with graduate programs in 1961. More briskly than in academia, systems engineering grew in industry. Its first hotbed was the aerospace industry. To develop the intercontinental ballistic missile (ICBM), the U.S. Air Force restructured its procurement process in 1954.[46] Instead of relying on an airframe producer as its prime contractor, it hired the upstart Ramo-Woodbridge Corporation (later TRW) for "systems integration and technical direction," effectively for overseeing and coordinating the ICBM development process. Electrical engineer Simon Ramo became an ardent proponent of systems engineering, which he defined as "a technique for the application of a scientific approach to complex problems. It concentrates on the analysis and design of the *whole*, as distinct from the components or the parts. It insists on looking at a problem *in its entirety*, taking into account all the facets and all the variables, and relating the social and technical aspects."[47]

Wartime urgency for weapons systems accentuated the importance of time and hence another central idea of systems engineering, concurrency. Its aim is to shorten the time required to bring a product from the drawing board to service by anticipating problems in manufacturing and operation and developing the product and its production process in parallel. Although the practice existed earlier, the term *concurrent engineering* was coined in the 1950s by Bernard Schriever, a mechanical engineer who directed the United States' crash ICBM program.[48]

The comprehensive design and development of an air defense system required expertise in many areas. Mathematicians and communication engineers at Bell Labs collaborated with aerodynamicists and mechanical engineers at other companies. More permanent organizations soon emerged, as Ramo recalled: "We have begun to develop good 'systems teams,' combines of individuals who have specialized variously in mathematics, physics, chemistry and many branches of engineering, sociology, economics, political science, physiology, business finance, education, and government. The systems team knows how to pool its knowledge and attack a problem from many directions."[49] Multidisciplinary teamwork has become a hallmark of systems engineering.

With its emphasis on the whole, teamwork encompassing diversity, and analysis of objectives, systems engineering spread. As the cold war wound down, commercial competition heated up. In the recent whirlwind of globalization, product development assumes the speed of a blitzkrieg. As the complexity of technological systems soars and their socioeconomic impacts escalate, the margin of error in decision making shrinks. To wrestle with them, the role of systems engineering expands, and so does its capacity, aided by information processing technologies.

Systems engineering demands the anticipation of many factors and exacts heavy penalties for mistakes. Initially it appealed only to those willing to pay dearly for time, and its early history is littered with botched projects. As its technology developed, however, its costs dropped compared with its benefits. Goaded by similar practices in Japan that gave that nation a competitive edge, and absorbing Japanese insight, systems engineering made its way into the U.S. automobile and electronics industries and beyond. The Defense Advanced Research Projects Agency launched a five-year program in 1988 that brought together industries and universities to develop concurrent engineering methods. Now concurrency has become a mark of excellence in rapid commercial product development.[50]

The U.S. Department of Defense has been issuing systems engineering guidelines for decades. Major professional engineering societies joined in in the late 1980s. At the end of the twentieth century, MIT consolidated systems engineering, operations research, technology and policy, and related programs and established its Engineering Systems Division. A similar consolidation at Stanford created the Department of Management Science and Engineering. Both have similar missions, to employ systems engineering in complex design problems to serve sustainable development of the information-intensive society.

Software Engineering: Close Kin

Systems engineering overlaps considerably with software engineering. Software is a computer-based system. It can stand alone, as in airline reservations, or be embedded as a subsystem in a larger system, as in an aircraft autopilot. Software engineering involves many techniques peculiar to programming, for instance object-oriented technology. However, it shares with systems engineering many general principles for managing complexity, including requirements specification, abstraction, modularization, and life-cycle planning.[51]

"Code and fix" is a most popular approach in programming. Programmers begin with some general idea of the software's intended functions, which may or may not have formal specifications. Then they use whatever familiar techniques they have to write a little code, run it, and modify it if it does not work properly. This approach can achieve a lot, especially in the hands of sharp and seasoned hackers. For large and complex systems, however, it often leads to costly confusion, delay, and even debacle. The operating system for the IBM 360, introduced in 1964, took twice the scheduled time to finish. Brooks, its project leader, summarized his harrowing experience in *The Mythical Man-Month*. The cover of this classic on software development features mammoths struggling in a tar pit, representing the difficulties engineers encounter in struggling with software complexity.

Moore's law—the density of computer chips doubles every eighteen months—is well known. Economists have been scratching their heads over why the fierce growth in computer power has produced only mild growth in industrial productivity. Many factors contribute to the productivity puzzle. One was suggested by Microsoft's chief technology officer Nathan Mythrvold. He observed that Microsoft Word version 1.0, introduced in 1983, had 27,000 lines of code and version 95 had about 2 million lines. The same trend obtained for Basic and other Microsoft products. Citing these examples, he concluded: "So we have increased the size and complexity of software even faster than Moore's law. In fact, this is why there is a market for faster processors—software people have always consumed capability as fast or faster than the chip people could make it available."[52] The growing obesity of software, however, is not nearly matched by rising software performance. Perhaps this can be called, for the sake of symmetry in the Wintel (Windows-Intel) duopoly on personal computers, the Gates law. Software developers can afford to be wasteful because hardware engineering is so successful in making computing power almost a free commodity. Software bucks the general engineering principle of economy, as computer scientist Jim Horning lamented: "Despite a half-century of practice, a distressingly large portion of today's software is over budget, behind schedule, bloated, and buggy."[53]

Problems with software were acknowledged early on. In 1968, in an effort to address them the Science Committee of the North Atlantic Treaty Organization invited top experts for a conference entitled Software Engineering. The IEEE defines *software engineering* as "the application of a systematic, disciplined, quantifiable approach to the development, operation, and maintenance of software." It divides into two areas, according to software engineer B. W. Boehm. The first consists of detailed design and coding, with little consideration for economics; the second involves requirements analysis, design, testing, and maintenance of application software in an economics-driven context. Scientific foundations are much stronger in the first area than the second.

Despite great improvements in development processes, the software crisis has persisted. The public is not amused by the large amount of money wasted on failed attempts to modernize the federal air traffic control system, the Internal Revenue Service's computers, and other software-intensive systems. Nor are fiascos limited to the public sector. In 1995 in the United States a total of about $250 billion was spent on software, and much of it was wasted. The private sector alone sank roughly $81 billion into software projects that were eventually abandoned.[54]

Software engineering is difficult, not least because it lacks the discipline of physical laws. As Brooks remarked, it is "pure thought stuff, infinitely malleable," "invisible and unvisualizable." Moreover, it is caught up in the "Red Queen effect." In Lewis Carroll's *Through the Looking Glass,* the Red Queen told Alice, "In this place, it takes all the running you can do, to keep in the same place." In a place where the enabling ground moves at the speed of Moore's law and the demand ground moves at the speed of information processing for globalization, one can hardly blame software engineers for panting and puffing. Software engineering still has a long way to go, but no one can deny that, relative to fixed ground, it has made tremendous strides.

Comprehensive and Long-Term View of an Engineered System

A system, even one as sprawling as a telephone network, is a bounded unit. Depending on the permeability of their boundaries, systems can be closed or open. Many scientific theories address closed systems, looking at their internal structures and ignoring their relations to the outside. Engineers, more interested in open systems that interact with a larger environment, define a technological system's internal structure together with its input and output. Even when input and output are not explicitly stated, they are hidden in the notion of *functions:* a system performs certain functions to serve the outside world. Extending the view of the system along its functions to the people it serves, one arrives at the comprehensive and long-term horizon of systems engineering.

The interpenetration of systems engineering's organizational and physical technologies is illustrated in Fig. 5.3. On the organizational side, systems engineers convene two large groups of people for multidisciplinary teamwork: the system's clients and workers. *Clients* broadly include sponsors who finance, owners who demand profit, users who expect service, licensing officials, regulatory agencies, environmental advocates, and those who may be affected by side effects, good or bad. *Workers* include engineers, scientists, manufacturers, suppliers, operators, maintenance crews, and others who contribute to developing and running the system. Each party has its own claims on the proposed system. Some may be unclear about what they want, others fight over conflicting interests. To find out what the system is supposed to do, what it is designed for, systems engineers must listen to various clients and help them to clarify their wants. After analyzing the demands, verifying their feasibility, and resolving conflicts, engineers synthesize the refined demands into a coherent set of requirements for the proposed system.

On the physical side, engineers consider the system's structural composition and temporal duration. For structure, they employ the systems

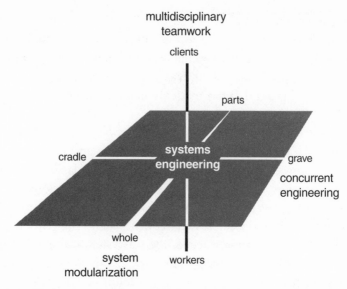

Figure 5.3   The temporal, structural, and organizational dimensions of systems engineering.

approach to design effective modules and interfaces. As designers of the system, they will work on it only for a short time at the beginning of its long life. Nevertheless, their thinking must cover all phases of its life cycle, from development through production and operation to decommission. They must consider how to produce, operate, maintain, and finally dispose of it, and design the needs of the *how* into *what* the system will be. Thus they make sure that their system design is economical to manufacture, easy to maintain, friendly to users, and benign to the environment. This up-front consideration of a system from cradle to grave is the gist of *concurrency* in systems engineering. It requires input from myriad experts, leading to the organization of cross-functional teams.

5.3 "Working Together" in Aircraft Development

Systems engineering, which creates complex systems by integrating physical and organizational technologies, grew rapidly and spread widely in the last decades of the twentieth century. "The Boeing Commercial Airplane Group is using it to develop the giant 777 transport and expects to release design drawings a year and a half earlier than happened with the 767. John Deere & Co. used it to cut 30 percent off the cost of developing new construction equipment and 60 percent off development time. AT&T adopted it and halved the time needed to make a 5ESS electronic switching system," reported *IEEE Spectrum*, which also provided detailed case histories from Hewlett-Packard and Cisco Systems.[55]

Development of the F-117A Stealth Fighter and the 777 Transport

For two aircraft development projects that exemplified good systems engineering, let us look at the F-117A stealth fighter by Lockheed (now Lockheed-Martin) and the 777 transport by Boeing Commercial. One emphasized product and the other, process.[56]

Lockheed Advanced Development Projects is a special projects unit that was established in 1943 by renowned aircraft designer Kelly Johnson. Its "engineers were young and high-spirited, who thought nothing of working out of a phone booth, if necessary, as long as they were designing and building airplanes," recalled one of the group.[57] Working out of a stinking facility, they called it the Skunk Works and turned the division into a model emulated by, among others, the Phantom Works of McDonald Douglas (later bought by Boeing, which would compete with and lose to Lockheed-Martin in the bid to develop the F-35 strike fighter for the twenty-first century). The chief mission of Skunk Works was technological innovation: to develop, prototype, and produce limited quantities of new models with relatively high speed and low cost. Its low overhead enabled it to take higher than usual risks in technology. Risks entail failures, and Skunk Works had its share, but they were outshone by its successes. Systems engineers Kevin Forsberg and Harold Mooz wrote: "Knowledgeable people have raved about the success of the Lockheed Skunk Works, which in 1945 produced America's first operational jet fighter (the P-80), and in 1962 produced the world's fastest airplane (the SR-71 [Blackbird], which held the world speed record for 30 years), with a project cycle time less than half of what others required." An important reason for its success, they continued, was that "Johnson was a perceptive and intuitive systems engineer."[58] Two of its products still in service are the U2 spy plane and the F-117A Night Hawk, developed in 1979 under Ben Rich, who had succeeded Johnson four years earlier.

In January 1991, when the F-117A flew in Operation Desert Storm to destroy heavily defended Iraqi command centers and make the sky safer for conventional aircraft, the 777 was three months into its full-scale development. "Boeing is developing the 777 transport in an atmosphere of sweeping changes in the way it designs and builds aircraft and in its corporate attitude toward customers, suppliers, and its own employees," reported *Aviation Week*.[59] "It represents systems engineering on a grand scale," remarked systems engineer Paul Gartz, who participated in the work.[60]

Besides engineering excellence, Skunk Works and Boeing Commercial have little in common. Skunk Works had a small core, consisting of an intensely cohesive group of some fifty veteran engineers and twice as many expert machinists. Its products were top secret, officially non-existent even when they flew missions. In developing the F-117A, it scored a breakthrough in stealth technology by making the aircraft as inconspicuous to radar as a tiny ball bearing. This high innovation in product design, however, involved little innovation in the design process, which had already been streamlined in traditional Skunk Works systems engineering.

Boeing's commercial arm is a giant that engaged some four thousand engineers in the 777 development project alone. Its products are advertised when they exist only on paper. The 777, the latest in Boeing's familiar line of passenger jets, incorporates many new technologies, including composite structural materials and fly-by-wire, which augments the pilot with sophisticated electronic flight control systems. However, Philip Condit, chief engineer of 777 development before becoming chief executive of Boeing, remarked that while these technologies were big advancements, they were not breakthroughs. Most had already appeared in Airbus products, not to mention military aircraft. The breakthrough for the 777 was not in product design but in the process of its development. Condit explained that improvements in commercial aircraft performance, such as range and speed, have reached the point of diminishing returns. Therefore the industrial emphasis has shifted to lowering cost and price, which demands superior engineering and manufacturing processes. He wrote: "I believe the fundamental orientation of the organization will shift and become process-oriented as opposed to product-oriented."[61]

Process innovation demands as much technology as product innovation. The Boeing 777 is lauded as a "paperless" airplane because its design was completely defined digitally, reducing or eliminating costly drawings and mockups. The Catia computer-aided design and computer-aided manufacturing systems tied together hundreds of design teams, ensured consistent definitions, and helped to reduce changes by more than 50 percent. Condit cited the Catia as a technological breakthrough in process engineering, but added: "Computers don't design and build airplanes—people do."[62]

Concept Definition: Getting It Right at the Beginning

An engineering design process can be divided into two major phases: *conception* and *full-scale development*. In the conception phase, engineers work with intended clients to determine the system concept, which includes a set of functional requirements that the system must satisfy and a definition of the system's architecture by which it performs the functions. If the proposed concept is approved by the clients, the system enters full-scale development, in which its detailed structures are determined.[63]

Speed, not of the aircraft but of its development, is one of the most acclaimed achievements of systems engineering. At first glance, speed is not apparent in the time lines of the F-117A and 777, as shown in Fig. 5.4. On the whole, they appear to have been prolonged projects. The crux, however, lies in the development phase, which for the F-117A lasted only a year before production began. The development phase of the 777 was about twenty-four months, compared with forty months for Boeing's previous model, the 767. Counting from the initial concept for the new aircraft, however, the 777 was slow to get rolling. Its concept took four years to get approval, compared with less than a year for the 747. But the haste of the 747's early phase exacted a heavy price when jumbo airframes later rolled off the assembly line with concrete blocks in place of jet engines under their wings. That fleet of sitting 747s, waiting for engines and bleeding interest payments, dragged Boeing to the brink of bankruptcy and Seattle into a recession. In contrast,

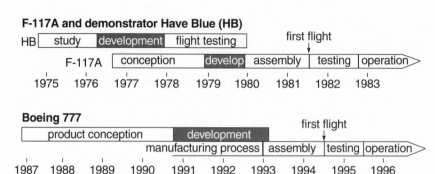

Figure 5.4   Development timelines for the F-117A stealth fighter and 777 transport.

the 777 not only met schedules but was service-ready on the day of delivery. The difference was in the wisdom of systems engineering: Take time to think it through before making a move, so that when initiated the move is swift and strong.

The U.S. Air Force, not known for its patience, was not rash in the case of the F-117A. It had been shocked when, in the 1973 Yon Kippur War, Israelis flying American planes with the best electronic countermeasures were shot down easily by Arabs using Soviet radar-guided guns and missiles. To avoid such casualties, ways had to be found to neutralize the radar defense. Among the options explored, a top candidate was a strike aircraft that could penetrate the radar screen undetected. Despite the urgent need, the Air Force did not rush blindly into development. It asked DARPA to undertake a year-long study, which concluded that stealth technology would be feasible but difficult. Physical laws dictate that to reduce the range of radar detection by a factor of ten, the aircraft's radar cross section must be reduced by a factor of 10,000. To minimize risks in developing the stealth fighter, the Air Force decided to commission two two-thirds scale demonstration units to test the new technology. Code-named Have Blue, the relatively inexpensive demonstrators helped engineers solve many problems that would have plagued a hasty full-scale development with exorbitant costs and delays.

What does one gain from early study and concept definition, which were substantial for both the 777 and the F-117A? Study costs less than development and much less than production. Because interest payments are often a significant part of a project's cost, rapid development and production are desirable even without considering the first-comer advantage in marketing. Furthermore, changes are inevitable in designing a system of any complexity. The further into the development process one gets, the harder it is to make changes. When more details are fixed and components are more interrelated, a change has more ramifications.

Careless changes can be disastrous, as the Apollo 13 accident showed. The temperature controller of Apollo 13's cryogenic oxygen tanks had been designed for operation at 28 volts. Later changes in the spacecraft design required the controller to operate at 65 volts. This simple change in requirements entailed extensive changes in the controller, which had many components. While making the modifications, engineers overlooked the thermostat switch. The mistake escaped several levels of review and made its way into the Apollo 13 spacecraft. Combined with a series of minor accidents, the thermostat's malfunction caused an oxygen tank to blow up during a routine stir, ruining the mission.[64]

The Apollo 13 case may be extreme, but changes in design are generally costly and risky, late changes especially so. Alan Mulally, who succeeded Condit as head of 777 development, compared changes in detailed designs with wielding a chain saw inside an airplane, ripping up structures to replace a component. "No more chain saws" was a slogan of his.[65] To avoid late changes, things must be done right at the beginning, preferably the first time. This implies emphasizing up-front work and a longer period for conception and gestation. It usually pays off. An Air Force study found that spending more on conceptual design can reduce revisions and total project time by up to 40 percent. Similarly, a NASA study found that cost overruns drop precipitously for projects that spend more than 5 percent of development costs in the conception phase.[66]

What Are We Designing For? Requirements Elicitation

Product conception typically consists of two periods, *requirements definition* and *system definition*. The two are most clearly divided in government procurement. The Air Force wanted a stealth aircraft. After feasibility studies, it defined a set of requirements, which included reduction in radar observability and other performance characteristics needed to meet military objectives. Then it requested proposals for demonstration units that would satisfy the requirements. Lockheed, Northrop, and other interested contractors each researched radar-absorbing materials, aircraft shaping, and specialized avionics, then optimized trade-offs to define a system—an aircraft design—and submitted it as a proposal. After reviewing the proposals, the Air Force chose Lockheed to develop Have Blue. Thus requirements definition and systems definition are undertaken by two groups in the public procurement process. They are often performed by one body—a corporation—in the private sector, but are punctuated by trips to the boardroom, where engineers seek approval for their proposal and resources for systems definition.[67]

"What exactly do you want?" is a question as common in engineering as in daily life. Customers order a technological system for certain objectives. When they are unsure of or unrealistic about what they want, engineers work with them to analyze their situation and clarify their desires. An objective can usually be met in many ways. Engineers draw on their knowledge about the industry, consult regulations, survey enabling technology, perform relevant research, estimate costs and schedules, assess risks and benefits, make trade-offs, suggest options, and confirm that the options do advance the objective. Sometimes the objective has to be modified to fit available resources or other constraints. Iteration between the objective and the functions it requires yield increasingly refined and satisfactory requirements. The process, called *requirements engineering,* is often the joint effort of engineers and clients.

A set of requirements specifies an engineering goal to be met. We saw how Black discovered that changing his goal from preventing distortion in amplifiers to canceling amplifier distortions made the difference between failure and success in inventing negative feedback. Goals have an equally strong influence on solutions in large projects. If the requirements are badly or wrongly stated, the designs are unlikely to be good, no matter how hard the engineers try. Summing up his thirty years of experience in designing aircraft, Johnson observed: "The most common cause of failure is lack of a suitable engine when needed. The second most common cause is working to the wrong (or constantly changing) set of requirements."[68] Other engineers might put inadequate requirements above inadequate enabling technology. Brooks wrote: "The hardest single part of building a software system is deciding precisely what to build. No other part of the conceptual work is as difficult as establishing the detailed technical requirements . . . No other part of the work so cripples the resulting system if done wrong. No other part is as difficult to rectify later."[69]

Horror stories about inadequate requirements abound, especially for software and information processing systems, where users are more prone to be unsure of what they want. A recent study by the International Electrochemical Commission identified poor specification to be the root cause of 44 percent of failures in safety related control systems.[70] Inadequate requirements can also kill engineering projects. A high-profile case in point is the Advanced Automation System (AAS), the centerpiece of the ambitious program at the Federal Aviation Administration (FAA) to modernize the nation's air traffic control system. The FAA's main objective was to equip air traffic controllers with new displays and computers for processing radar and flight data. After a four-year design competition, the FAA awarded the development contract to IBM, to which it handed a thick book of requirements. As IBM's work progressed, however, the FAA came up with an unending stream of changes. As the pile of requirements documents grew from four feet to twenty feet tall and extensive codes had to be rewritten, the relationship between the FAA and IBM went from being collaborative to adversarial, which is deadly for any engineering project. Schedules slipped, costs mounted, Congress fumed. After ten years, the FAA abandoned the AAS in 1994, wasting an estimated $1.5 billion in development costs.[71]

Requirements engineering worked well in the development of both the F-117A and the 777. The primary military objective for the stealth fighter was to perform surgical strikes against heavily defended strategic targets with minimal loss of aircraft. The Air Force resisted temptations to add secondary objectives such as air-to-air combat, and frills that could have added enough complications to jeopardize the primary objective. To transform the military objective into aircraft requirements, DARPA and the Air Force analyzed combat data and performed engagement simulations to find the radar signatures that would provide satisfactory invulnerability under realistic battlefield conditions. These studies set the requirements for the Have Blue demonstrators. Test flights of Have Blue improved knowledge and yielded the requirements for the F-117A. Besides trading off range, speed, weapons, and other operational capabilities, the requirements study also balanced performance with the risk of pushing the technological envelope and the time required to reach initial operation. Unlike what it did in the development of some other weapons systems, here the Air Force decided not to push too many technologies at once. In analyzing the F-117A development program, aeronautic engineers David Aronstein and Albert Piccirillo attributed much of its success to its "simple and realistic" requirements: "Narrow focus—achieve a very low observable system quickly and secretly—was the key to the program success. All other performance goals were traded off when necessary to preserve the primary program objectives."[72]

As the customer and user of the F-117A, the Air Force knew what it wanted. Boeing was the developer and seller of the 777, not its user. It keeps abreast of transportations needs, market conditions, fuel prices, regulations, airfield facilities, and other industrial conditions, but this general information, also available to its competitors, does not reveal the specific interests of individual customers, which it must woo. Boeing did not beat formidable competitors to become the world leader in commercial aircraft by being insular. "The single most important element of success is to listen closely to what the customer perceives as his requirements and to have the will and ability to be responsive," said Jack Steiner, chief engineer for developing the Boeing 727 in 1960, about his project.[73] Few lessons were more conspicuous in the product. When it was conceived, the 727 had two prospective buyers. Eastern Airlines wanted an airplane with two engines; United wanted one with four. A compromise was struck at three. A two-sentence story necessarily misses all the intricate negotiations leading to agreement, which Steiner identified as among the most difficult tasks. And that is not to say that technical design was easy. The trijet was an odd and unpopular configuration. Just to position the three engines was a headache for which Steiner commissioned competing teams. But his respect for customers paid off. With 1,831 delivered, the 727 became one of Boeing's best-selling airliners, second only to the 737.

Condit went further to create an organization in which the 777 developers could listen not only closely but aggressively to customers, in more depth and regarding details of design. He knew from experience that engineers usually try hard to understand what they believe the customers want. However, because the structure of many large corporations gives design engineers little access to customers, they can only conjecture based on their limited knowledge. Consequently their efforts can be less than effective. Here is where organizational technology can make a big difference in improving the process of development. Project leaders can reorganize the work environment by bringing customers into the room, so that design engineers can stop guessing and obtain information directly from the customers. Thus, starting from conception, Boeing made it a policy to invite airlines and airplane leasing companies to join in the design process, and the invitation was extended throughout development. This was an innovation in the 777 development process.[74]

How Well Does It Perform? Requirements Verification

A requirement, no matter how desirable, only fattens litigation lawyers if the product cannot be tested for compliance. A good set of requirements includes appropriate methods of verification. Toward this end, engineers introduce various performance indices and metrics. This is not always easy, especially when the requirements involve human factors. Nevertheless, they try. A good example is seen in assessing flight-handling quality.[75]

When Air Force pilots first saw the F-117A, they exchanged anxious looks and said: "Boy, it sure is an angular son of a bitch." Totally constructed of flat angular plates for low radar observability, the aircraft so violated good aerodynamics that many thought it would not fly, even less maneuver like a fighter. Rich assured the pilots that the alien-looking bird would be "most responsive" and "easy to handle."[76] Responsiveness and good handling sounded too vague to be verifiable, but the engineer was not making empty promises to the pilots whose lives would be entrusted to his product. An aircraft's handling quality determines the ease and precision with which its pilot can perform the tasks required to accomplish his mission. Providing good handling quality is an objective of flight control systems, especially for military aircraft that require high maneuverability. Verifiable specifications for handling quality are difficult to determine because they entail quantifying the pilot's subjective experience under various conditions, including a varying workload. Research to tackle this problem, initiated after World War I and intensified under the pressure of World War II, proceeded together with studies on flight stability and control. Many types of aircraft were tested, flight simulators were developed, and theoretical models were introduced that included representations of the pilot in analyzing pilot-aircraft control loops. Working together, pilots and engineers developed standardized definitions and rating scales to debrief test pilots, quantify their opinions, and correlate them to measurable quantities such as stick force per g and other flight performance. By 1942, the Air Force was able to issue handling-quality specifications for its aircraft. Since then, the requirements have been continually refined and modified for new classes of aircraft.

Human-Machine Interface

Technological systems are designed by people to work with people for people. Therefore the human-machine interface is one of their most important aspects. Inadequacy in this interface has ruined many products. Take the Advanced Automation System project, for example. The FAA had decided that the federal air traffic control system could not be totally automatic but had to involve human controllers, whom the AAS was intended to aid but not replace. However, it hardly consulted the controllers themselves when stipulating the requirements for the AAS. When IBM built a simulation center to study how its design worked for the controllers, it was buried by demands for changes and was stunned to find much of its design useless to the controllers. It is to prevent such fiascoes that systems engineering states loudly: Involve all stakeholders, including operators of the intended system, early in conception and requirements definition.

Proper human-machine interfaces are difficult to design, especially for new technologies for which users do not know what to expect. Eliciting their responses is a complicated task for engineers. Working out a way to specify flight handling qualities took more than two decades. For this and many other human-factor requirements, many techniques of elicitation have been developed, often with the help of psychologists and other social scientists. For the 777's flight deck, Boeing solicited early design suggestions from airlines, pilot associations, the FAA's flight-test branches, and human-factors experts. It also hosted joint reviews for engineers and pilots.

A successful technique for requirements elicitation is to build a demonstrator or mock-up like IBM's simulation center for air traffic controllers. Before the real working mechanism of a system is developed, a demonstrator simulates its interface with its users. Interacting with it helps users to clarify what they want. Studying human-machine interactions and performance results, engineers learn what is wanted and what to improve. Flight simulators are common; aircraft manufacturers routinely build mock-ups of their airplanes to help both their engineers and their customers. The Boeing 777 team eliminated physical mock-ups except for the nose cone, because its Catia computer system simulated everything digitally. The nose cone mock-up verified that computer-aided design had worked satisfactorily. Among its jobs was to check the computer's performance in designing ergonomics. Catia generated digital models of human beings so that design engineers could ensure that inspection and maintenance crews would have easy access to various parts of the aircraft.[77]

Concurrent Engineering as Working Together

Concurrent engineers take a long-term view and reckon with a system's needs and performance throughout its life span. For this they consider its *dispositional* properties in addition to its *substantive* ones. Substantive properties, such as weight and speed, are a system's characteristics at particular times. Dispositional properties, such as solubility and flammability, are how the system will or will likely behave under certain conditions. These properties are dealt with in natural science but receive much more attention in concurrent engineering, which strives for verifiability, manufacturability, usability, reliability, maintainability, interoperability, flexibility, portability, evolvability, reusability, disposability, and a host of other abilities. Dispositional properties are less tangible than substantive properties and are sometimes defined only in terms of probability. To anticipate relevant situations and frame clear concepts for dispositional properties, perhaps to quantify them, are tasks for concurrent engineering.

Anticipating future contingencies, concurrent engineers call on workers involved in different phases of the system's life cycle to form multidisciplinary design teams. Among future phases, manufacturing and operation are most important. Operation usually belongs to the customers' organizations: airlines for Boeing and the Air Force for Skunk Works. Boeing, which had traditionally listened to the airlines' business needs, went further in its 777 development to take account of the technical details of operation. In designing for maintenance, Boeing engineers invited airline maintenance crews into the process, learned about service and repair, and designed ease of maintenance into every part of the 777. Because a lower maintenance cost for the airlines may mean a higher manufacturing cost for Boeing, trade-offs had to be worked out. Taking the trouble to account for operational needs at the level of detailed design is a part of what Boeing calls its Working To-gether Initiative. Gartz explained: "Working Together was a Boeing and airline term for concurrent engineering. The program brought to-gether teams consisting of engineering, key airlines, suppliers, manu-facturing, and customer service organizations all at the front end of the program to define the requirements and designs."[78]

Without singing about it, *working together* was built into the orga-nization of Skunk Works from the beginning and its consistent practice contributed enormously to its success. "There were no prima donnas or glory hounds," Aronstein and Piccirillo reported. Instead, the "people involved developed a strong sense of teamwork and a high degree of motivation. The participants enforced a multidisciplinary, systems en-gineering approach." Engineers in charge of design or specialized in aerodynamics and other specific topics met constantly, conferring and kibitzing every step of the way. They also worked closely with test pilots, machine shop workers, and managers when they encountered problems. Two of the basic rules Johnson set for Skunk Works were that engineers should be "within a stone's throw" of the machine shop floor and that "bad news should travel to the top at least as quickly as good news." These rules were so ingrained that design engineers spent at least a third of their day right on the shop floor. At the same time, there were usually two or three shop workers up in the design room conferring on particular problems. The close-knit culture kept adminis-trative bureaucracy to a minimum and all workers involved and integrated on a project. It discouraged passing the buck and motivated workers to be supercritical of their work before they sent it on. Rich noted: "Self-checking was a Skunk Works concept now in wide use in Japanese industry and called by them Total Quality Management."[79]

"Communicate, communicate, communicate was instilled and worked," both within Skunk Works and between its workers and its customers. Throughout development and testing of the F-117A, Skunk Works engineers teamed up with Air Force aviators and service crews, who shared their personal experiences with off-the-shelf parts and ways to deploy maintenance vans, ensuring that the aircraft could be serviced effectively. James Allen, who commanded the F-117A unit when it achieved initial operational capability, estimated that "the joint validation/verification teams reduced the process time by half."[80]

Intensive communication between design and manufacturing facilitates a second sense of concurrency that is more risky and controversial. Ordinary concurrent engineering allows temporal overlaps in the development of a product and its production process, as in designing an aircraft and its production line simultaneously. In some desperately urgent cases, concurrency goes one step further and overlaps product development with actual production, as in starting up the production line when parts of the aircraft are still on the drawing board. The U.S. Air Force has been criticized for "flying blind" when it practiced such desperate concurrency. To achieve faster deployment, it risked waste and cost overruns, especially when a weapons system demanded ambitious technologies. Whether the risk was worth taking depended on national security considerations that are beyond our scope here. But when a political decision has been made that the risk of wasting money is preferable to the risk of wasting lives in war, managing the financial risk need not be flying blind, given sound concurrent engineering. Engineers can try to manage the unavoidable risk and reduce its effects, as they did with the F-117A.

The Air Force wanted an operational squadron of stealth fighters so badly it suspended its own rule of "fly it before you buy it." Before test flights of Have Blue were completed, and bypassing the usual process of calling for competitive proposals, it gave Skunk Works the contract to go ahead on full-scale development of the F-117A. Furthermore, the contract stipulated that production of the fighter should start in a year, before its prototype would be ready for flight testing. Both sides knew that the compressed schedule implied a risk of costly modifications to production aircraft when test flights revealed problems—and problems were inevitable, given the airplane's radically new technology.

Skunk Works took risks but was not reckless. It carefully scheduled Have Blue flight tests so that it got the vital data first. Knowing that they could not avoid retrofits and modifications in production models, engineers designed for this from the start. Rich explained: "We had planned for concurrency from the beginning by keeping detailed parts records on all the production models and designing easy access to all onboard avionics and flight control systems." These procedures enabled them to use successive production aircraft as "guinea pigs" for testing aerodynamics and propulsion, because changes and improvements could be easily retrofitted into earlier products. Rich concluded: "The bottom line in concurrency is cost savings, provided it is done right."[81]

Scaling Up Concurrency

What makes a group effective is often called its "chemistry." Just as chemical reactions resist straightforward scaling from test tubes to industrial tanks, workplace chemistry changes beyond recognition when the number of workers increases ten or a hundredfold. Skunk Works' F-117A development team consisted of a few hundred workers. When Boeing was of a similar size in its early days, it too was tightly knit. The entire design group worked within fifty feet of each other on the floor above the factory, and engineering and manufacturing went back and forth. Looking back on company history, Condit remarked: "As the scale goes up, that gets harder and harder and harder."[82]

As scales expand with mass production, work organizations tend to fragment. Engineering and manufacturing departments may be located in different cities or even different countries. The intensive division of labor means that various phases of a product's life cycle are assigned to different groups of highly specialized workers who scarcely talk to each other. Workers become shortsighted. Engineers create a design and throw the blueprint to the manufacturing department, sometimes only to have it thrown back with a nasty comment about the specified configuration being impossible to make. This "throw it over the wall" mentality, which results in expensive reworking and redesign, has worried engineers and managers in many industries since the beginning of the twentieth century, and many attempts have been made to redress it. Steiner saw it as a culprit in the substantial manufacturing cost overrun of the Boeing 707. Determined not to repeat the mistake in his development of the 727, he tried to get manufacturing's signatures of approval on engineering's drawings. When the manufacturing department refused to accept the responsibility, he hired a group of industrial engineers and manufacturing personnel and put them in the center of the development team to review every layout before final drawings were made. This worked well. However, Condit was not satisfied: "At Boeing, we can trace our past successes to the fact that we listened closely to what our customers want of an airplane, but our future rests on also listening as a manufacturer."[83]

How to bring thousands of workers in various functions together and encourage them to communicate, communicate, communicate? Information technology succeeds in greatly diminishing the "here and there" barrier caused by physical distance, but it is less effective against the "us and them" barrier caused by social organization. Condit's solution for lowering the psychological barrier was to organize the 777 development workforce into design-build teams: "You break the airplane down and bring Manufacturing, Tooling, Planning, Engineering, Finance, and Materiel all together in that little group. And they are effectively doing what those old design organizations did on their bit of the airplane."[84] Cross-functional design-build teams made Boeing's Working Together Initiative effective. At its peak, 238 teams were active, each working on a part or a subsystem of the 777.

The design-build teams concept has won wide acclaim. Its effectiveness is premised on two central ideas in systems engineering. First, a complex system, such as an airplane, can be modularized or broken down into subsystems whose detailed designs are more tractable. Second, a subsystem is treated has having a *general* structure similar to that of the system as a whole, operating on a smaller scale. The processes of requirements definition and systems definition occur on a lower level. Designers of a subsystem, say the details of an aircraft's wing, also have to work closely with customers and capture their requirements, although their customers are not the airlines but the engineers designing the aircraft's overall structure. They, too, have to get manufacturing approval for their designs, and their manufacturers have more specialized knowledge. Thus the small-group dynamic, which worked so well in Skunk Works for designing the F-117A, was reproduced in each design-build team responsible for a small part of the 777. Modularization and similarity of organization on a smaller scale are central to the top-down systems design approach.

5.4 From Onboard Computers to Door Hinges

Through requirements elicitation and system conception, engineers sketch out the overall characteristics of a system in broad strokes. The system definition serves as the top-level framework for development, but it is only conceptual. To transform it into a physical product with concrete details that work, engineers generally practice what Brooks called "top-down design" for complex software, which identifies "design as a sequence of *refinement steps.*"[85] It is not unlike a painter sketching a whole layout on a huge canvas, then filling in the details layer by layer. The procedure is reasonable but seems to contradict one of the most popular distinctions between science and engineering, that scientists analyze and engineers synthesize. Intuitively, synthesis is a bottom-up approach that combines elements. How is it compatible with top-down design?

Iterative Analysis-Synthesis-Evaluation Cycles

Via analysis and decomposition, science proceeds from organisms to organs to cells to genetic molecules to nucleotides to atoms to nuclei to elementary particles. Engineers put things together, synthesizing numerous components in creating, say, an airplane. Different as they seem, the two activities are not mutually exclusive but cooperative. Analysis and modularization are indispensable in engineering, while scientists synthesize many causes in explaining a complex natural phenomenon. The symbiosis of analysis and synthesis, which has deep roots, is an outstanding feature of the systems approach.[86]

Socrates subscribed to the "methods of division and collection."[87] Similar methods were widely practiced in the Renaissance. Leonardo da Vinci's notebooks contain not only remarkable machine designs but

also analyses of forces and systematic studies of gears and other machine parts. Perhaps most famous is his anatomy, which is an analytic
study of the human body. Michelangelo also took pains in dissection,
even at the price of falling ill from handling corpses. Leonardo explained: "It is a necessary thing for the painter, in order to be able to
fashion the limbs correctly in the position and actions which they can
represent in the nude, to know the anatomy of the sinews, bones, muscles, and tendons in order to know in the various different movements
and impulses which sinew or muscle is the cause of each movement,
and to make only these prominent and thickened."[88] Similarly, to capture the vitality of the living body in marble, the sculptor analyzes the
individual joints and muscles and examines how they work together in
the whole body. These artists' approach to anatomy shows that analysis is not necessarily looking at parts individually without regard for
the whole. Leonardo and Michelangelo examined bones and muscles
from the perspective of the live human body in which they function.
Their approach was at once analytic and synthetic. Engineers call it the
systems view.

Galileo's methods were often described as consisting of resolution
and composition.[89] They became analysis and synthesis in the writings
of Newton and René Descartes. Newton described the methods in both
mathematics and natural philosophy: "By this way of analysis we may
proceed from compounds to ingredients and from motions to the forces
producing them, and in general from effects to their causes and from
particular causes to more general ones, till the argument end in the
most general. This is the method of analysis; and the synthesis consists
in assuming the causes discovered and established as principles, and by
them explaining the phenomena proceeding from them and proving the
explanation."[90] Descartes wrote: "We first reduce complicated and obscure propositions step by step to simpler ones, and then, starting with
the intuition of the simplest ones of all, try to ascend through the same
steps to a knowledge of all the rest."[91]

While other methodologists only juxtaposed analysis and synthesis,
Descartes expressly combined them into a single rule of thought, writing, "This one Rule covers the most essential points in the whole of human endeavor." Analysis is conducted not for its own sake but for solving some complex problem or explaining some phenomenon, for which

analytic results are judiciously synthesized. In this context, *synthesis* has two roles. First, it offers a synthetic perspective for the problem or phenomenon as a whole, a synthetic conceptual framework in which analysis is to be conducted. Second, it refers to the integration of analytic results. Thus analysis and synthesis form a round-trip from the whole to the parts and back, from top to bottom and back. For instance, atomic physicists analyze an atom into the nucleus and "shells" of electrons and scrutinize their behavior in detail, but they ultimately synthesize the subatomic components into the structures of various atoms as wholes.

The analysis-synthesis loop is often iterated. Scientific analysis is often approximate. In Plato's metaphor, scientists try to carve nature at its weak joints, taking care not to mangle it like an unskilled butcher. Nevertheless, to make headway they sometimes sever joints that are not so weak. Results of crude analysis are synthesized into approximate solutions, which are tested by experiments and improved by more refined analysis. Thus researchers go back and forth between the whole and its parts, successively clarifying them. Iterative analysis-synthesis-evaluation loops are also the crux of systems engineering. Design engineers move back and forth between the requirements of a system as a whole and the detailed design of its components.

From System Definition to System Architecture

Cognitive scientist Herbert Simon told a parable in his *Science of the Artificial*. A watchmaker designs a watch with a hundred parts that must be assembled in one breath. Because the process is excruciatingly intricate and any interruption or mishap ruins the whole thing, its cost is high, resulting in an expensive watch. His competitor designs a watch with similar performance, except that it is made up of ten modules, each with ten parts. Because a defective part affects only one module and not the whole watch, the process of assembling a module or combining the modules is relatively simple and inexpensive. He can offer his watches at a lower price and easily beat the first watchmaker in business.

Simon's two watch designs illustrate two interpretations of a system as a whole with interrelated parts: the *seamless web*, popular in armchair ideologies, and the *system* of practical engineering. Like the first watch, a seamless web is a whole that must be modified as a whole. Although it sounds profound, seamlessness is ill-suited to handle complexity, for its resistance to analysis breeds muddled concepts and dogmatic thinking. Even worse, it is prone to disaster, because the effects of a minor local flaw easily propagate, triggering breakdowns elsewhere, and causing web-wide collapse. Sensible designers of practical systems strive to avoid this predicament by introducing seams that stop undesired unraveling but are not absolutely impermeable. Like the second watchmaker, they break the whole into modules, which facilitates scientific analysis and piecemeal improvements. Systematic modularity enables you to replace a broken part in your car quickly, seamlessness would demand rebuilding the whole car. An engineering system is known not for its seamlessness but its good seams. The Internet, for instance, is designed so that its seams, the routers, connect disparate networks while preserving the autonomy of those networks' internal operations.

Modularization, in which a complex system is broken down into modules or subsystems with proper interfaces, is the crux of engineering's systems approach. Vincenti explained this, using detailed cases from aeronautics. He delineated four levels after project definition: overall design, major component design, subdivision of component design according to engineering disciplines, further division for detailed design. "Such successive division resolves the airplane problem into smaller manageable subproblems, each of which can be attacked in semi-isolation. The complete design process then goes on iteratively, up and down and horizontally throughout the hierarchy."[92] Proceeding from overall definition to substantial details is the mark of top-down design.

When a system to be developed exists only as a concept, design engineers tend to analyze it according to the *functions* that the modules will perform. Many functional units turn out to have clear physical boundaries, such as the wings for lift and the fuselage for payload in the classic airplane configuration. But this coincidence is not necessary, lift and payload are distributed in the integrated airplane configurations, and lack of physical boundaries is the rule in software. In any case, the primary design concerns are functions. Because functions are high-level characteristics of a system, they accentuate the synthetic flavor of engineering analysis.

Functional modularization turns the system definition into *system architecture*. Just as scientists try to carve nature at its weak joints, engineers try to fashion modules with *strong internal cohesion* and *weak external interaction*. Weak coupling is not no coupling. The modules must work with each other, and their functions must combine to produce satisfactory system performance. Thus the interface between modules must be carefully designed to allow maximum independence for the modules, connecting them only as necessary and sufficient for the function of the whole. Too little interaction may deprive a module of the input it needs to perform its tasks, too much may create interference. Good interfaces are ubiquitous in engineering systems.

In the Boeing 777, propulsion accounts for about 7 percent of costs, and the airplane structure and aerodynamics 33 percent. The bulk of the cost, some 54 percent, goes to aircraft systems: electrical, hydraulic, and pneumatic equipment; environmental controls; flight control and autopilot; navigation, communication, and other avionics. The 777 incorporates sixty-six systems. And this is only the system- or aircraft-level modularization. The airframe was divided into the wings, fuselage, and tail. A wing performs so many aerodynamic tasks that its leading and trailing edges were divided functionally. The trailing edge was again modularized into spoilers, ailerons, various kinds of flaps and supports. The detailed design of the trailing edge alone called for ten design-build teams, each with ten to twenty engineers.[93]

Functional Decomposition and Physical Integration

The systems approach is aptly represented in the V model by Forsberg and Mooz (Fig. 5.5).[94] The downstroke of the V expresses top-down design, in which system characteristics are increasingly fleshed out in subsystems of smaller scales and greater detail. The upstroke expresses the implementation of designs and the integration of subsystems. In the system architecture obtained in top-level modularization, each subsystem is assigned certain functions that it must contribute to the system. The allotted functions can be viewed as the *objective* for the subsystem. To subsystem designers, the situation is similar to that of system design-

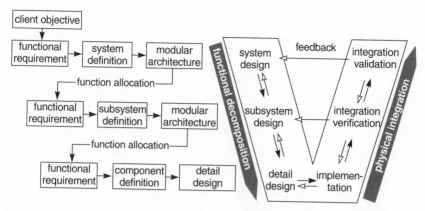

Figure 5.5 Forsberg and Moorz's V model of systems engineering highlights system modularization and integration. Each level of decomposition has its own design requirements, with which the finished subsystems must be tested for compliance.

ers, except that the "client" with the objective is now not some outside party but the system architecture. To transform the objective into a definite design, the process of requirements specification starts all over again.

As an example, let's look at the Aircraft Information Management System (AIMS), one of the sixty-six aircraft systems in the 777, developed by Honeywell working together with Boeing. Running 620,000 lines of code, which amounts to more than a third of the 777's non-entertainment software, AIMS is the largest computer on board. Its objective was to act as a platform that integrates and provides shared computing resources to several avionic functions: displays, flight and thrust management, flight data acquisition, data conversion and communication, flight deck communication, airplane condition monitoring, and more. To do its job, the AIMS had to interface with other major avionic systems and "talk" to the rest of the 777. The interfaces were part of its requirements, the specifications for which fell mainly to Boeing engineers who reckoned with the whole aircraft. But just as Boeing worked closely with the airlines to define requirements for the 777, Honeywell worked closely with Boeing on the AIMS requirements and was free to challenge its demands. Boeing requested that the 777's software be written in Ada, a programming language designed for embedded software but considered by some contractors to be "immature." Honeywell accepted the requirement only after conducting a study on the relative merits of Ada versus C for the project. For the initial requirements push, sixty-five Honeywell engineers went to Seattle for three weeks. To facilitate system integration and verification, engineers in the two companies established a special electronic link and held daily teleconferences. When the danger of endless requirements expansion loomed, an agreement was reached to allow a change only if it met some stringent criteria that affected the 777 on a wide scale.[95]

In AIMS development, Honeywell practiced as much systems engineering as Boeing. It modularized the functions of AIMS into seven major groups, each assigned to a team of up to a hundred software engineers. Robust software partitioning was instituted to ensure the functional independence of these software groups, so that they would not adversely affect each other's operations. The modular architecture allowed software packages to share processor, operating system, memory, data bus, and other resources. It saved power, reduced hardware weight, and increased reliability and maintainability. The AIMS design process entailed the same cycle of requirements capture, system definition, modularization, and functional allocation as the aircraft as a whole.

The some 20,000 parts of the 777 were manufactured by 241 companies in 12 countries. Although made to precise specifications, each had to be tested and integrated into subsystems that would work for the airplane. For this purpose Boeing built the Integrated Aircraft Systems Laboratory (IASL), which at its peak employed 1,500 workers, mostly engineers. On delivery, components were checked and assembled into the appropriate subsystems. These were tested in one of eight subsystem integration facilities, one of which was dedicated to AIMS. After passing individual tests, subsystems moved to three major integration facilities, where they were tested for their interaction with other subsystems under simulated flight conditions. In a facility equipped with actual 777 avionics, aircraft-level electrical connections, and high-power computers providing real-time simulation of the aircraft in flight, AIMS was "flown" through various conditions by test pilots who provided evaluation and feedback.[96]

The IASL was planned during the conception of 777 development, under the rationale of concurrent engineering. By testing aircraft systems extensively before the aircraft's first flight, the amount of actual flight testing was reduced. Accelerated testing enabled the 777 to enter service already certified for extended-range twin-engine operation that allowed it up to three hours' flying time from an airport. This was an unprecedented feat, partly made possible by integrated ground-testing technology.

Systems Thinking on the Individual Level

Systems engineering is most well known in the development of large-scale projects. As design thinking writ large, however, it is also useful for developing systems on smaller scales, just as debates on the justice of public policies help us to understand the justice of personal actions. A similar process of requirement-definition-architecture was used in designing the 777 and AIMS—and may be used generally for a system and its subsystems—although the two are on different scales. This does not mean that design organizations think of scale linearly, so that a large project is just more of the same. The scaling of design processes is not linear but akin to what mathematicians call a fractal.

A *fractal* is something that exhibits self-similarity at different scales. A familiar example is a rugged coastline. Photographs of a coastline taken from a satellite and from an aircraft show similar ruggedness, although they are on very different scales. Attempts to calculate coastal length from satellite photographs, taking proper account of magnification factors, will yield an underestimation, because more capes and coves show up in close-up views, adding more length. It is not simply a matter of scale but a peculiar *combination of scale and resolution,* such that a similar general pattern shows up on all scales with their corresponding resolutions.

The fractal idea can be borrowed to illustrate how systems engineering treats not only scale but a *combination of scale and proper abstraction.* Abstraction, which hides irrelevant information, is important in systems engineering and in science. The top-level view takes in the full scale of a system, but it is rather abstract. As one descends to the smaller scales of subsystems, more concrete details are added. With proper trade-offs between scale and concreteness, a general pattern of design thinking obtains on different levels.

Consider, for example, the detailed design of the 777's passenger doors, especially the door hinges. A hinge is a small piece, but engineers did not jump right into designing it. The design-build team for passenger doors started by analyzing the records for the Boeing 767 and discovered that in the design process for its doors, 13,341 changes were made at a cost of about $64 million. Determined to cut that cost by at least half, they decided that the key to a better design was for all the door hinges to share common parts. This had not been done previously because the curvature of a fuselage requires that each of the eight doors has its unique hinge. To design hinges with common parts for disparate doors demanded inventive thinking. The best engineer on the team accepted the challenge, and after three months of hard work and near despair, he succeeded. Moreover, the tricks he discovered in wrestling with the hinges were generalized to design common parts for 98 percent of all door mechanisms. Looking only at the hinges, designing eight sets of parts for eight different hinges would certainly be easier than designing a single set of parts common to all eight. But the requirement of common parts stemmed from a broader, systems concern: it would save cost in production and design changes. The engineer had to look beyond the hinges to the larger goal.[97]

By emphasizing objectives, systems thinking encourages engineers to break free of conventional thinking and look at more issues. Many engineers are experts in optimizing the functions of specific devices. With a broader systems vision, they go beyond local optimization and think more about global optimization, and thus discover more design possibilities.

Rational Organization as Enabler of Creativity

The systems engineering organization is often caricatured as hierarchical, undemocratic, and hostile to creativity. A system composed of parts that are composed of even smaller parts is a hierarchy. Let us for argument's sake also call the organization of system designers a hierarchy, although engineers prefer to see it as a partnership among groups working on parts large and small. The question is, how much is the organizational hierarchy one of social control and power domination, and is the control more stifling than in other work relationships?

A work organization that is known to drive out talent and strangle creativity is micromanagement; few like to have the boss breathing down their neck, especially a pompous boss ignorant about what is really going on. An ideology of micromanagement is the postmodern insistence on the *thorough* social construction of all scientific facts, so that cultural ideologies determine not only the agenda of research projects but also the content of research results. The postmodern misrepresentation of actual research practice has been much criticized.[98]

Systems modularity helps to prevent top-down design from falling into micromanagement through the clear division of responsibilities. Rich continued the Skunk Works tradition when he became its leader: "I'll be decisive in telling you what I want, then I'll step out of your way and let you do it." This attitude is more than a personal leadership style. It has been formalized in the process of requirements definition, which sets clear boundaries of responsibility and allows free rein of creativity within those bounds. In working out the specifications for the AIMS, Boeing engineers reported that "one of the key contributions of Honeywell's participation was to ensure Boeing stayed focused on defining only necessary system requirements and not implementations." Both sides recognized the advantage of "requirements independent of implementation." It curbed the reach of Boeing's top-down design and prevented interference with Honeywell's inventiveness in seeking the best way to meet the requirements.[99] Honeywell was below Boeing, if one insists on calling the systems organization a hierarchy, but it was not socially controlled by Boeing beyond the usual work relationship that anyone doing a job for someone else accepts. Organizations can be stifling, as can anarchies or cultural norms that force conformity. However, they are not necessarily so. Just as a social contract enables citizens to enjoy more freedom than in an anarchy by agreeing on certain laws of government and obeying them, a systems organization can be an enabler of creativity.

"There is no single best method to develop an aircraft" was the first important lesson offered by Johnson.[100] Similarly, Brooks's slogan "No silver bullet" is widely accepted, even as software engineers struggle to find methods with superior effectiveness.[101] Systems engineering is no silver bullet, but in coordinating creative talent to design complex systems, it seems superior to many more idealistic doctrines.

6

SCIENCES OF USEFUL SYSTEMS

6.1 Mathematics in Engineering and Science

Besides the know-how vitally embedded in human expertise and physically embodied in machines, engineering possesses a large store of explicitly articulated knowledge. This includes more than huge databases on physical and structural parameters that are invaluable in making design choices; engineering has long advanced beyond the stage of cataloging facts. A 1955 report by the American Society for Engineering Education (ASEE) identified engineering sciences aside from the basic sciences of mathematics, physics, and chemistry. Some engineering sciences, such as chemistry, are specialized to particular branches of engineering. Others are common to many branches. Among these, the ASEE report named six: mechanics of solids, fluid mechanics, thermodynamics, transport phenomena, electromagnetism, material structures and properties. And it added with emphasis: "*It may be anticipated that other engineering sciences will develop;* for example, information theory shows promise of contributing to measurement and control in all engineering fields."[1]

An engineering science delineates the underlying principles and mechanisms for a type of existent or possible artificial system with physical bases susceptible to deliberate design and control, for instance systems that utilize heat. Engineering shares with physics the laws of thermodynamics, but differs from it in the vast amount of detail regarding how thermal processes work in various classes of heat engines, chemical reactors, and other useful systems. Like natural sciences, engineering sciences are deliberately explained and organized for effective communication and education. They constitute a significant portion of undergraduate curricula and a much larger portion of graduate research in engineering schools. Today's technological systems are so complex someone ignorant of them would be an incompetent engineer.

Engineering sciences roughly fall into two classes: *physical* and *systems*. The former examines topics according to underlying mechanisms; the latter, their functions. "Physical" refers broadly to material-based phenomena, including the chemical and the biological. With the rapid advance of bioengineering, cellular or molecular biology may well emerge as a new engineering science. Systems sciences are more indigenous to engineering. Three major groups of systems theories cluster around control, communication, and computation.

Mathematical Representations of Reality

Leonardo wrote: "Let no man who is not a mathematician read the elements of my work."[2] Galileo said that the great book of the universe "is written in the language of mathematics."[3] "The unreasonable effectiveness of mathematics in the natural sciences," as physicist Eugene Wigner put it, has puzzled many.[4] Equally puzzling is the effectiveness of mathematics in engineering. Why is differential geometry, far more abstract than Euclidean geometry, more suitable for representing spatiotemporal structures? Why is systems theory, closer to pure mathematics than mathematical physics, important to real-life engineering problems? Is Einstein perhaps right that the human mind conceives forms before finding them in things, so that in creating complex forms, mathematics helps scientists to descry subtle structures in nature and engineers to design them?

Visualization, a good way to conceive spatial forms such as buildings, is highly valued by engineers and scientists, as is apparent in the efforts to develop graphic data presentations. However, constrained by the characteristics of our sensual organs, it is limited to a small range of forms. Many physical properties and processes, from quantum mechanics to electronic signals, are not conducive to visualization. For

complex and nonspatial forms, mathematics, with its capacity to articulate relational structures, is perhaps the best medium for conception.

Aristotle counted mathematics as a *téchnē,* an art. Ancient mathematics addressed mainly tangible things and methods useful in commercial reckoning. It stood out among contemporary arts and sciences
for its clarity and precision in long chains of deductive thinking. As rigorous thinking developed and expanded far beyond immediate experience and utility, it grew increasingly abstract and asserted its autonomy. A turning point in this long ascent of abstract thinking occurred
in the seventeenth century. Alongside the scientific revolution, and crucial to its success, modern mathematics took shape with the introduction of analytic geometry by René Descartes and Pierre de Fermat and
calculus by Gottfried Wilhelm Leibniz and Isaac Newton. Concomitantly, Western philosophy shifted its focus from metaphysics to epistemology, from God to the nature of human thinking.

An often-cited advantage of mathematics in science is its ability to
provide quantitative descriptions. Instead of qualitative adjectives that
are often crude and vague, the number systems provide countless predi-
cates that are refined, precise, general, and versatile. Quantification
facilitates measurement, which is more accurate than sensual experience for empirical verification, more public, and less susceptible to variations among individual persons. This stimulates the introduction of
objective concepts. All of this is conducive to science, but the power of
mathematics goes much farther. When we say that A is 10 meters long
and B is 5, we not only describe the properties of A and B separately
but also tacitly state that A is longer than B. Furthermore, we can add
the two to find their combined length. Constituting a system with intricate structures that support arithmetic operations, numbers are good
for representing not only properties but also relationships and procedures.

Descartes recalled that he was unimpressed when he first studied
arithmetic and Euclidean geometry; they bogged the mind down with
numbers and figures, which seemed hardly worth the intellectual labor.
Then he realized that the exclusive concern of mathematics is *relations*
and *patterns of relations.* "Relation" is a generic notion covering everything from operations and transformations to the temporal relations
among successive stages of a process. And it is irrelevant whether what

are related are numbers, shapes, stars, or anything else. Just as Galileo separated "machine in the abstract" from "machine in the concrete," Descartes abstracted mathematics from figures and numbers. He found a powerful method for achieving this in algebra, which he integrated with geometry to create analytic geometry. His account of his thinking processes brought out several central features of mathematics: symbolism, abstraction, generalization, rigor, relational structure.

Words in natural languages are symbols rich in nuance but prone to misunderstanding, for their meanings are loaded and vague. To avoid excess meaning, mathematical symbols are specially designed to concisely embody concepts that are precisely defined and clearly related. Good symbolism is instrumental to complicated and accurate reasoning, the superiority of Arabic to Roman numerals being a case in point. Symbolism also facilitates analysis and synthesis. Just as engineers combat complexity by introducing modules to hide details not immediately useful, mathematicians introduce a symbol to represent a cluster of concepts, thereby hiding the cluster's internal details and accentuating its external relations to other concepts. In this way, wrote Descartes, "nothing might needlessly occupy our mental powers when our mind is having to take in many things at once."[5]

Descartes compared three representations of the right-angle triangle. The arithmetic representation, which uses a group of numbers such as (3, 4, 5) for its sides, obscures its being right-angled by the lumped description of the hypotenuse. Euclidean geometers represent it by the figure for the Pythagorean theorem: a triangle the lengths of whose sides are related by the areas of squares sitting on them. It is clumsy and difficult to generalize and manipulate. The algebraic representation in symbolic form $(x, y, \sqrt{(x^2+y^2)})$ brings the relation among the three sides of the triangle distinctly into view, reveals the essence of a right-angle triangle, covers countless cases, and is susceptible to generalization and manipulation. Its superiority is apparent.

Euclidean geometers reckon with length, area, and volume, which are distinct magnitudes not susceptible to generalization. In contrast, the algebraic representation of x, x^2, and x^3 suggests the relation of multiplication, which can be readily generalized to x^n, x multiplied n times. This opens enormous room for conceptual maneuvering, for instance allowing combinations of all powers of x to represent points on a curve. Symbolic representation gives geometry a new life with unlimited growth potential.

Descartes, who introduced the symbol x^n, valued it only for its clarity and economy. Pierre-Simon Laplace, who came decades later, recognized the fruits of its susceptibility to extension by analogy. Once the symbol of integer exponents x^n for powers took hold, others quickly extended it to fractional exponents $x^{n/m}$ for roots and subjected them to similar manipulations. Citing this example, Laplace observed: "Such is the advantage of a well-constructed language that these most simple notations have often become the source of most profound theories."[6]

Mathematics can be regarded as a language designed for the creation, manipulation, and analysis of elaborate concepts, conceptual relations, and conceptual structures. It ensures that a concept has precise meaning within its conceptual structure, and its relations to other concepts, however strenuous, can be traced clearly. Its austerity unclutters the mind, concentrating and amplifying conceptual creativity. Such a precise language is invaluable for engineering and natural science, which demand imagination restrained by clear reasoning to fathom complex structures of the real world. If a realistic situation can be represented adequately in the language of mathematics, engineers and scientists can view it from various mental perspectives, connect the results from one perspective to another, bring out its factors in forms most perspicuous, zoom in for minute details of particulars or out for big pictures and their generalities, make approximations with clear knowledge of their significance—all with the high security provided by sufficient rigor. The *if* of this statement is not assured, but numerous successes in science and engineering have demonstrated that it works in many important cases.

Abstraction and Relational Structures

Problems with two unknowns representable algebraically, as $2x + y = 3$, existed in antiquity but were ignored for the lack of determinate solutions. Then Descartes and Fermat explicitly introduced the concept of *varying magnitudes* or *variables*. At once the indeterminate equation leapt to life as representing the *relation* between the two variables x and y. Thus $y = 3 - 2x$ represents a relation that traces a straight line as the variable x takes on various values. It is susceptible to graphic representation and, more specifically, becomes a part of analytic geometry. Furthermore, $y = a - 2x$ represents a class of straight lines that are parallel to each other; $y = a + bx$, all straight lines. This simple example illustrates two points: abstraction and levels of generality.

Mathematical abstraction does not so much discard inessential features as bring out essential features so forcefully that the rest is eclipsed. A variable abstracts from the specifics of all the values it can assume by hiding them under a single symbol to bring out formal relations. Once we grasp the general form, we can substitute any value for the symbols x and y to generate results for special cases. Generality is conducive to representations of scientific theories, a major strength of which is counterfactual inference that tells us what would happen if something was the case.

Assigning values to some but not all variables offers views on different *levels of generality:* all straight lines, a class of lines satisfying a general criterion, or a particular line. We customarily distinguish *variables* designated by x and y from *parameters* designated by a and b. Fixing a parameter, such as $b = 2$, narrows the focus to a specific class for its details. By varying the parameters, we trade detail for scope and thus proceed to a higher level of generality that covers many cases. Like an intellectual lens that can zoom from the microscopic to the telescopic, mathematics enables scientists to shift their perspective and look at their subjects in various levels of detail.

The notion of variables was vaguely known before Descartes and Fermat. In making it explicit and drawing out its logical consequences, they opened the door to endless conceptual constructions that could be based on it. Leibniz called the relation between variables a *function,* which has proliferated to all corners of mathematics and acquired the precise definition of a *mapping.* A function f is a rule that systematically maps a value of x, called its domain, into a unique value of $y = f(x)$, its range, as in Fig. 6.1a. The general concept of a mapping between a domain and a range becomes the platform for other abstract constructions.

Looking at a Complex System Many Ways

The concept of a mapping from a domain to a range places no restriction on the nature of the domain or range. Under the general definition

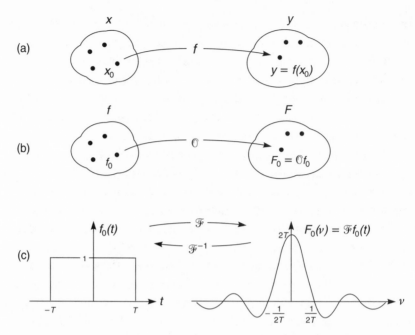

Figure 6.1   (a) A function *f* is a mapping that assigns a unique element y_0 in its range *y* to an element x_0 in its domain *x*. (b) Generalizing the idea, an operator $\mathcal{O}$ is a mapping between a domain *f* and a range *F*, the elements of both being functions. (c) An example of an operator is the Fourier transform $\mathcal{F}$ that systematically maps a function *f* with time *t* as variable to a function *F* with frequency *v* as variable. In the specific case illustrated, the Fourier transform reveals that the frequency bandwidth of a square pulse with time duration 2*T* is inversely proportional to *T*.

of a mapping, the function becomes a special case in which the domain and range consist of real or complex numbers. Because the mathematical crux lies in the *mapping,* nothing prevents the objects mapped from being something more complicated. The objects can be functions themselves, in which case we get *operators,* such as $F = \mathcal{O}f$, where the operator $\mathcal{O}$ relates two classes of functions *f* and *F* by mapping a function f_0 in the domain *f* to a unique function F_0 in the range *F* (Fig. 6.1b). Operators, which relate functions, are examples of high-level abstract construction. They find extensive employment in science and engineering. Two examples are the Fourier and Laplace transforms.

Many dynamic physical systems are represented by a function *f*(*t*) governed by a differential equation, where the independent variable t represents time. The Fourier transform $F(v) = \mathscr{F}f(t)$ is an operator $\mathscr{F}$ that maps functions $f(t)$ into functions $F(v)$ with variable v, physically interpreted as frequency. Because of the physical interpretations, $f(t)$ is called the *time domain* representation of the system, whereas $F(v)$ is the *frequency domain* representation. When the "frequency" variable is complex, $s = \sigma + iv$, the transformation $F(s) = \mathscr{L}f(t)$ becomes the Laplace transform. The real part σ of s is associated with gain and loss, hence the Laplace transform is useful in analyzing system stability.

Time and frequency are both important variables for a physical system. By connecting them, the Fourier transform reveals essential characteristics of the system. Figure 6.1c illustrates a simple case: a square pulse in the time domain has finite energy only during the time interval $2T$. Its Fourier transform in the frequency domain reveals that most energy of the pulse falls within a frequency bandwidth that is inversely proportional to T. This relation explains the advantage of broadband in telecommunications. Consider transmitting a data stream, a sequence of 1s and 0s where 1 is represented by a square electrical pulse and 0 by the absence thereof. A high data rate implies many pulses per unit time, which in turn demands short duration for each pulse. And a short pulse, the Fourier transform tells us, corresponds to a large frequency bandwidth.

The time and frequency domains related by the Fourier transform are *two representations* of *one system*. The importance of looking at a complex problem or physical system from many perspectives is stressed by many engineers and scientists. Changing the scientific perspective is not changing the topic, a trick politicians use to evade difficult questions. Researchers stick to a complex problem and probe it from every angle possible. Sometimes a problem that is hopelessly opaque and difficult in one perspective becomes clear and easy when viewed from another perspective. By transforming a problem, mathematics can greatly simplify it. For instance, multiplication is a complicated procedure, but is simplified by taking the logarithm, which reduces multiplication to addition. Similarly, the Fourier and Laplace transforms simplify the job of solving many differential and integral equations by reducing them to algebraic manipulations. Where they are applicable, they enable engineers to characterize and analyze many systems in simple forms that highlight their physical and functional significances.

Mathematical Algorithms for Computation

Electronic computers have changed all areas of engineering and science by facilitating solutions of equations, simulating realistic situations, automating experiments, processing signals, storing large amounts of data, presenting data to aid visualization, and mining data for useful patterns. Although popular images of computation are dominated by hacking and the raw power of computing hardware, mathematical reasoning is necessary for successful computation.

Much of computational science and engineering involves numerical solutions of physical laws and other differential equations, many of which cannot be solved analytically except in the simplest configurations. Many interesting situations are so complex even the most powerful computer cannot handle all their details. Therefore conceptual abstraction and mathematical modeling are indispensable. After a model is formulated, mathematical reasoning continues, for instance in analyzing discretization and rounding errors in computation. Furthermore, mathematical insights have produced many algorithms that perform computational tasks efficiently. They account for most of the top ten algorithms that the journal *Computing in Science and Engineering* identified as being most influential in the development and practice of science and engineering.[7] For instance Monte Carlo methods rely on the central limit theorem in the probability calculus to compute problems with high dimensions. The algorithm of fast Fourier transform is another example.

Physical signals in telecommunication and other applications are usually measured in the time domain, where their amplitudes change. However, they are processed more effectively in the frequency domain where, for instance, a filter can be designed to eliminate high- or low-frequency noise. Casting the measured signal into the frequency domain for processing requires the Fourier transform, and casting the processed signal back for transmission, the inverse Fourier transform. Suppose a digital signal within a certain time interval is mathematically represented by a vector with n elements, each representing the signal at a particular moment. To cast the signal into the frequency domain, the discrete Fourier transform (DFT) multiplies it by an $n \times n$ matrix. In straightforward matrix calculus, the computation involves roughly n^2 numerical multiplications. Now $n \approx 80,000$ for a 10-second digitized telephone conversation, and its DFT involves about 6 billion multiplications. The number is much larger for music, image, and other complicated signals. Furthermore, the transformation must be performed in real time, taxing the power of the computer.

James Cooley noticed in 1965 that the DFT matrix can be factored in such a way as to minimize the number of multiplications. As a result, the number of numerical multiplications required for Fourier transform is only of the order of $n\log n$. This mathematical algorithm, implemented in a computer program by John Tukey, is the fast Fourier transform (FFT). Fast it is, requiring less than one ten-thousandth the time to process a 10-second speech signal with ordinary DFT. Valuable for telecommunication and digital signal processing of all kinds, it is used everywhere, from radar and sonar to biomedical equipment.[8]

The crux of FFT lies in mathematically analyzing matrices and unscrambling the results. It is not tied to the electronic computer, although it increases computational efficiency. A similar idea was proposed by Carl Friedrich Gauss. To discern subtle patterns in complex systems is always a forte of mathematics, which can contribute elsewhere as to computational science and engineering.

Mathematical Rigor and Physical Reasoning

Physical phenomena obey natural laws. Mathematical structures conform to the rigor of reason. Representing physical reality by mathematics, straddling the concrete and the abstract, scientists and engineers must carefully balance the two in their modeling and concept formation. Biases frequently result in complaints that mathematics occupies too much or gets too little attention, or that a gap opens between theory and practice. A historical story serves as a good introduction to this important point.

In his book on operator calculus, mathematician Harold Davis traced five movements in its historical development. The first movement included the works of Leibniz and Lagrange, the second Laplace, the fourth Fredholm and Volterra, the fifth Hilbert and Schmidt, all great mathematicians. In the middle he wrote: "Strangely enough, the third movement in the theory of operators was initiated by important researches in electrical communication. The protagonist of this dramatic story was Oliver Heaviside (1850–1925), a self-taught scientist, scorned by the mathematicians of his day, who saw only yawning chasms of unrigor behind his magic formulas."[9]

Heaviside came from a poor English family and educated himself working as an electrician. While other autodidactic engineers became entrepreneurs, he made himself into a research engineer when that was almost unheard of. He was one of the few who championed Maxwell's field theory when most physicists deemed it speculative. Stripping it of obscuring mechanical scaffolding, he supplied many details, introduced rationalized electromagnetic units, and wrote down the four Maxwell's equations in the modern form that we recognize today.[10] Applying the field theory to engineering problems, he developed a comprehensive theory of transmission lines and suggested several methods to obtain ideal distortion-free propagation. One of his methods was to insert induction coils, which, later developed and sold to AT&T by Pupin, got long-distance telephony off the ground. It brought half a million dollars to Pupin but nothing to Heaviside, who published his results instead of patenting them and died starving.[11] Yet Heaviside had the satisfaction of developing theoretical results and disseminating them among engineers. His mathematical method, although less than rigorous, indeed worked like magic in calculation. Its merits were acknowledged by mathematicians in the early twentieth century. In their efforts to justify and extend it, they found it similar to Laplace's "generatrix calculus" and called it the Laplace transform, and it has become an indefatigable workhorse in science and engineering.[12]

Heaviside's dispute with contemporary mathematicians and his vindication by later generations reveal the subtle relations between mathematics and physics. Heaviside worked in the spirit of Euler, Lagrange, Laplace, and other great mathematicians of the eighteenth century who made long strides in developing calculus. They forged ahead with their calculations, and by twentieth-century standards defined terms vaguely, handled convergence sloppily, generalized carelessly. Yet subsequent rigorous scrutiny found that they had committed few serious errors, partly because they were guided by the physical intuition they gained from applying their mathematics to solving problems in mechanics.

Since the nineteenth-century turn to higher abstraction, mathematics began to disengage from physics, as Georg Cantor declared in 1883: "Mathematics is entirely free in its development."[13] For freedom, mathematics pays the price of empirical intuition and support. Now pure abstract thinking can rely only on its own rigor for justification. Rigor becomes more vital as abstract construction produces structures with ever greater complexity, where slack thinking easily leads to errors.

Living in a time when mathematics was worrying about its own foundations, Heaviside's work was scorned by pure mathematicians for its lack of rigor. Working in applied mathematics, Heaviside fired back and argued that rigor should not be rigor mortis. His points were not totally outrageous among mathematicians. Felix Klein, a renowned pure mathematician of his age, wrote in a similar vein: "The ideal of 'rigor' has not always had the same significance for the development of our science . . . In times of great, powerful productivity it has often receded to the background in favor of the richest and quickest possible growth, only to be all the more stressed once again in a succeeding critical period, when the concern is to secure the treasure already won."[14]

To say that concepts in axiomatic mathematics are judged solely by their formal coherence means only that they are not bound by any physical or utilitarian applications, not that they are necessarily unsuitable for, cut off from, or unmotivated by applications. Mathematics and science continue to challenge and fertilize each other. Heaviside's impulse function and Dirac's delta function, which mathematicians declared nonexistent but engineers and physicists successfully used, motivated mathematicians to develop distribution theory for generalized functions. Conversely, the intellectual freedom in abstract constructions can facilitate invention and scientific discovery of the unknown. Liberation from prior experiences of the familiar can open the mind to esoteric phenomena or new designs. The dethroning of Euclid's parallel postulate opened the door to alternative geometries and Einstein's theory of general relativity. The theory of prime numbers, which seems to be only of academic interest, becomes a cornerstone for useful cryptography. For empirical science and engineering, higher mathematical abstraction is like a high-orbit satellite equipped with long lenses. Its lofty perspective enables it to see large and subtle structures of the earth invisible from the ground.

A rocket that soars too high can escape from earth and never return. Mathematical thinking produces so many conceptual structures of such subtlety it can be enchanting. To be enthralled by abstract mathematical beauty to the neglect of concrete problems is a danger for engineers and natural scientists. Heaviside saw it clearly: "The practice of eliminating the physics by reducing a problem to a purely mathematical exercise should be avoided as much as possible. The physics should be carried on right through, to give life and reality to the problem, and to obtain the great assistance which the physics gives to the mathematics."[15] Feynman's personal experience bears testimony to Heaviside's assertion that all good mathematical scientists "obtain the important assistance of physical guidance in the actual work of getting solutions." Feynman was proficient in mathematics, but he said he tended to think in terms of physical cases rather than mathematical generality. He told how he stunned a formal speaker by spotting mistakes in a complicated mathematical theory he was unfamiliar with: "He thinks I'm following the steps mathematically, but that's not what I'm doing. I have the specific, physical example of what he's trying to analyze, and I know from instinct and experience the properties of the thing. So when the equation says it should behave so-and-so, and I know that's the wrong way around, I jump up and say, 'Wait! There's a mistake!'"[16]

Mathematical Modeling and Representation of Reality

Mathematical science and engineering are about things and processes in the real world. The world is so vast and complex that any physical theory, no matter how sweeping, represents only a few facets of it, and even that with idealization and approximation. This point is emphasized in Einstein's letter that was discussed in section 5.1. Modeling, in which one delves into the tangle of real-world factors, picks out the essential ones, makes judicious approximations, and represents them conceptually, is a necessary step without which the most powerful mathematical machinery runs in vain. "Mathematical modeling does not make sense without defining, before making the model, what its use is and what problem it is intended to help to solve."[17] This assertion by biochemical engineer James Bailey is echoed by countless other engineers and scientists. Perhaps its most famous formulation, which applies also to computation, is also its most succinct: GIGO—garbage in, garbage out.

The same argument applies to engineering systems theories such as Shannon's information theory. Systems theorist Rudolf Kalman explained that the problem of control, in which engineers make a physical system perform as desired, falls into two parts. The first is model building, in which engineers frame a mathematical representation of the dynamic properties of the physical system to be controlled. The second, the locus of control theory, deals with "mathematical models of certain aspects of the real world; therefore the tools as well as results of control theory are mathematical."[18] Like other systems theories, control theory abstracts from most physical factors but not all. It retains essential *general* conditions. Thus the Kalman filter, a celebrated result in control theory, is a great improvement over the Wiener filter partly because it relaxes the latter's assumption of stationarity, which is not satisfied in most physical situations. With all their axioms and theorems and proofs, good systems theories have sunk their teeth deeply into engineering realities, where their ultimate significance lies. This was appreciated by mathematician Andrei Kolmogorov. Commenting on the success of information theory, he wrote that Shannon "combines a deep abstract mathematical thought with a broad and at the same time very concrete understanding of vital problems of technology."[19]

Heaviside wrote: "No mathematical purist could ever do the work involved in Maxwell's treatise." David Hilbert quipped that every schoolboy in the streets of Göttingen knew more about four-dimensional geometry than Einstein, but it was Einstein and not the mathematicians who discovered relativity. Mathematical science or engineering is like a solo concerto. Mathematical calculation takes up the most time, just as the orchestra produces the most sound. Calculation is guided by physical and design principles, just as the sonority of the orchestra is commanded by the virtuosity of the soloist. Important as mathematics is, in science and engineering, physical and design principles reign supreme. When they run dry or go awry, fancy formalism degenerates into what physicist Howard Georgi called "recreational mathematical theology." Engineers in MIT's Lincoln Laboratory put it most succinctly: Mathematics should be on tap, not on top.

6.2 Information and Control Theories

"There is nothing better than concrete cases for the morale of a mathematician. Some of these are to be found in mathematical physics and the closely related mathematical engineering," remarked mathematician Norbert Wiener, who contributed substantively to control theories. Quoting Wiener five decades later, electrical engineer Thomas Kailath wrote: "It must be said that the term 'Mathematical Engineering' does not enjoy the currency that the name 'Mathematical Physics' does. Being a younger field, its proponents still focus on more specialized descriptions such as Information Theory, Communication, Computation, Control, Signal Processing, Image Processing, *etc.* The names 'System Theory' or even 'Mathematical System Theory' have been advanced but are not universally accepted."[20]

All engineers and physicists use a variety of mathematics to address real-world problems. Mathematical physics and engineering employ more sophisticated mathematics, but neither is a unified theory. Engineering has many systems theories: linear, nonlinear, deterministic, stochastic, continuous, and discrete time theories; and the more topic-oriented information, control, computation, and signal-processing theories. They are more abstract than the physically oriented engineering sciences, and more mathematical in flavor, as control engineer Dennis Bernstein remarked: "Although other branches of engineering such as fluid mechanics and structural mechanics are major users of mathematics, none is as mathematical in style and spirit as is control. When was the last time you saw a theorem-proof format in a fluid or structural journal?"[21]

Bernstein also pointed to a paradox. Control theory is quintessentially engineering and practical. Despite its abstractness, it cries out for and is highly successful in hardware implementations. Control devices regulate lighting, elevators, and indoor climates; monitor security and fire safety; stabilize the swaying of tall buildings in high winds; adjust fuel injection, ignition timing, throttle, engine torque, and other functions in automobiles; fine tune temperatures, flows, and other variables in chemical processing plants. Without flight control, many high-agility aircraft, notably jet fighters, would be unstable and would crash in no time. These are only a few of the applications of control theory. They work so reliably in technological systems that we often entrust our safety to them without much thinking.

How does control theory, with its theorem-proof format, produce results that work meticulously in the real world? Why is mathematical rigor more prominent in systems than in physical theories? We will return to these epistemological questions about systems theories after we examine some specific examples.

The Rise of Systems Theoretic Thinking

Engineering theories before World War II were narrow in scope. Tackling servomechanisms, mechanical engineers developed theories mainly in the time domain, where the controlled plant was treated as a dynamic system. Tackling amplifiers, communication engineers developed theories mainly in the frequency domain, in which they solved problems of gain and stability. Then the war brought them together, especially in developing controllers for aiming anti-aircraft guns. When their minds met, they found that the time-domain and frequency-domain representations did not conflict but complemented each other. An electronic amplifier whose output had to track arbitrary input telephone signals with fidelity was in some ways analogous to a mechanical gun that had to track its target's erratic movements with accuracy. Articulating their similarities clearly led to systems theoretic thinking.

"The distinguishing characteristic of system theory is its generality and abstractness, its concern with the mathematical properties of systems and not their physical forms," electrical engineer Lotfi Zadeh explained.[22] Abstracting from most physical factors, retaining only general structural and functional characteristics of apparently disparate devices, and representing them by mathematical concepts, systems theories reveal the deep commonalities beneath various branches of engineering. In a systems theory, a low-pass filter is simply something that passes low and rejects high-frequency oscillations. It retains general physical characteristics, such as the cutoff frequency and insertion loss, but disregards most specifics, such as whether the filter is physically an electronic circuit or an automobile suspension. This idea of *systems ab-*

straction is illustrated by the lower row in Fig. 2.3 as compared with the upper row. Block diagrams similar to those in the lower row, which extract the relations among the components of a system, are common in engineering texts.

An engineering system is generally a set of interacting components subject to various inputs and producing various outputs. The system proper, without input and output, is usually a dynamic system whose state changes with time. Starting from the solar system, dynamic systems have been studied extensively at least since Newton. Modern dynamic theories, developed in the late nineteenth century, introduced many sophisticated concepts fruitful in both engineering and natural science.

Over the general notion of a system various complications can be introduced and constraints imposed. A linear system obeys the superposition principle. A feedback system's input is a function of the state of the system itself. Some systems are subject to haphazard disturbances, as ships under variable winds. Outputs of others are contaminated by extraneous factors, as when measurements are contaminated by noise or error. What conditions apply depend on reality and model building. Systems theories usually consider very general conditions, such as the whiteness of noise or the quadratic form of performance. Their abstractness enables their results to be implemented in a great variety of physical systems. A PID controller is generally defined by the proportional-integral-derivative form of its control law. Originally derived for steering ships, it has been generalized for use everywhere, for instance for controlling temperature and pressure in industrial processes.

Among the general concepts introduced into engineering during World War II, two, *optimization* and *probability*, have become cornerstones of systems theories. Optimization signifies the turn from piecemeal solutions to systematic rationalization of design. Mechanical engineer Ivan Getting recalled: "The practice before the war in the design of servos was to employ a mechanism *adequate* for the problem. The difficult problems encountered in the war, particularly in the field of fire control, emphasize the necessity of designing the *best possible* servo system consistent with a given kind of mechanism."[23] To assert what is the best is difficult, he continued. Engineers are too hardheaded to dream about the absolute and universal best, which anyone sensible knows to be infeasible. Optimization theories aim at the "best" relative to a specific objective criterion and a range of options and constraints. When these are formulated in a model that adequately captures essential realistic factors, mathematics provides many tools for deciding whether optima exist and if so, what they are. The mathematics of optimization, whose history goes back at least to the seventeenth-century calculus of variations, leaped forward in the 1950s.[24] Various mathematical programming (meaning planning) techniques provide powerful methods to obtain the optima for specific problems. They are widely used in engineering, operations research, and other areas.

The probability calculus that emerged in the early seventeenth century received a rigorous axiomatization in 1933 from Kolmogorov, who also applied probability to engineering problems. Independently of him and working with electrical engineer Julian Bigelow on the problem of filtering signals and predicting the aim of anti-aircraft guns, Wiener in 1941 argued for the probabilistic nature of communication. His monograph *Extrapolation, Interpolation, and Smoothing of Stationary Time Series* was called the "yellow peril" by engineers because of its yellow cover and difficult mathematics. It exerted a profound influence on the thinking of many engineers. Among them was Shannon.

Information Theory

Shannon's "A Mathematical Theory of Communication" inaugurated information theory, which provided the concepts, insights, and mathematical formulations that underlie modern communication systems. When the paper appeared in 1948, he had worked on it intermittently for eight years. Communication engineers to whom he referred had put forward many theories, including a logarithmic concept of information. However, they addressed the various components of a communication system piecemeal. No comprehensive conceptual framework existed that enabled engineers to integrate the components, relate the messages to be communicated and the signals that carry the messages, evaluate and compare the performances of competing technologies, and explore their potentials and limitations. Shannon's theory introduced such a framework. Communication engineer Robert Gallager explained: "It established a conceptual basis for both the individual parts and the whole of modern communication systems. It was an architectural view in the sense that it explained how the pieces fit into the overall space. It also devised the information measures to describe that space."[25]

"The fundamental problem of communication is that of reproducing at one point either exactly or approximately a message selected at another point," Shannon began in his paper. "We wish to consider certain general problems involving communication systems. To do this it is first necessary to represent the various elements involved as mathematical entities, suitably idealized from their physical counterparts."[26] His statement expresses the characteristics of a systems theory: mathematical representation, idealization from the physical, definition of a system in terms of its function, and simultaneous definition of its input and output, the source and user of messages.

Message sources produce voices, images, texts, data, and other physical signals. Communication channels occur in a great variety of physical media and are subject to noise that corrupts signals and introduces errors. Shannon delved beneath the diverse physics to seek their general characteristics pertinent to *reliable* communication—meaning that messages reach users with a vanishingly small probability of errors, despite channel noises, and without users' asking for clarification. He found the engineering significance of messages not in their meaning but in their being selected from a set of possibilities, not semantics but statistics. So he represented a message source as a random process whose statistical characteristics are summarized by a quantity defined as its *entropy H*. For the channel, he found the essential characteristic to be its *capacity C*, defined in terms of such general physical properties as bandwidth, noise spectral density, and signal-to-noise ratio. The entropy and capacity were introduced not independently but together with the signal that, carrying messages from the source through the channel to the user, ties everything in the system into a unitary whole. Design of signal forms is a major job of communication engineering. Shannon posed the general question: What is the best that signal processing can do to maximize the utilization of a noisy channel for reliable signal transmission? In other words, how far can engineers go to beat the noise, not by cranking up the transmitter power but with clever signal design?

He reasoned that messages in any representation, be it letters or

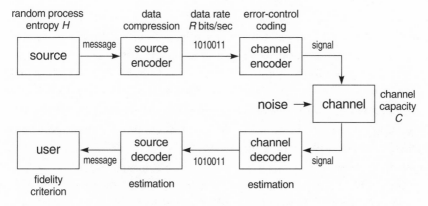

Figure 6.2   A communication system represented in information theory.

hieroglyphs, can be coded in a set of standard alphabets, the simplest of which are binary digits, or *bits,* with a data rate of R bits per second. The transmitter, which performs the signal processing and connects the source and channel, can be decomposed into two parts linked by the binary digits. The first part is the source encoder that conforms to the characteristics of the source, the second the channel encoder that conforms to the characteristics of the channel. A similar decomposition occurs on the receiving end (Fig. 6.2). The *separation principle* that decomposes the transmitter and receiver is another example of systems modularization. Practically, it makes the equipment simpler and more versatile. Theoretically, it underwrites Shannon's answer to his fundamental question.

Besides casting the message into binary digits, the source encoder compresses it by removing redundancy. Shannon proved in his *source coding theorem* that for a digital source with entropy H and information rate of r bits per second, its messages can be compressed into a binary sequence with average rate of R bits per second without losing information, so long as $R > Hr$. Since the value of H falls between 0 and 1, a significant amount of lossless data compression is possible.

The channel encoder transforms the string of binary digits into forms most suitable for transmission. Shannon's *noisy channel coding theorem* asserts: So long as the data rate R is smaller than the channel capacity C, it is possible to encode the signal at the channel input and decode it at the output in such a way that information is transmitted with a vanishingly small error rate despite the corruption of noise. In sum, reliable communication is possible so long as $Hr < R < C$. This result surprised engineers at that time because the data rate it predicted for reliable communication was much higher than they believed possible.

High transmission rates without error come with costs, among which are complicated and expensive signal-processing equipment. Nevertheless, channel coding is a novel way for ingenious signal designs to combat physical inevitabilities such as noise, limited power, or limited bandwidth. In designing error-controlling codes for certain types of channel noise and realistic parameters, engineers can call on the power of mathematics. The new dimension of maneuvering adds tremendously to engineering resources for achieving reliable communication.

Coding Theory

Newton's second law, force equals mass times acceleration, introduces the interrelated concepts of force, mass, and acceleration that form a general framework for nonrelativistic motions. Similarly, Shannon's theory introduces the interrelated concepts of source entropy, data rate, and channel capacity that form a general framework for communication systems. The definition of entropy is partly justified by its roles in the source- and channel-coding theorems that assert $Hr < R < C$ as the limiting condition for reliable communication. Newton's second law leaves open the form of the force, which is to be specified for each type of phenomenon, for example the inverse square law of gravitational force that Newton introduced separately. Shannon's theory leaves open the form of signal coding, for which other engineers developed coding theory.[27]

The channel coding theorem asserts the possibility for achieving arbitrarily high transmission accuracy in one-way communication, but Shannon's proof provides no clue about how to attain the performance. Coding theory, which aims to find practical error-control codes, quickly emerged as a substantial part of information theory. Its first applications were in deep-space communication serving the 1969 Viking Mars mission and the 1972 Pioneer Jupiter fly-by. Deep-space missions face stringent communication conditions and can afford expensive signal processing hardware, especially decoding equipment on Earth. A few other high-end applications existed, such as satellite communication. However, so long as equipment was expensive and leisurely transmission rates sufficient for mass traffic, no commercial pressure existed to push the theoretical limits. Information theory developed rather slowly. At its twenty-fifth anniversary, communication engineer John Pierce wrote: "What some of us attained was perhaps wisdom rather than knowledge."[28]

Even as Pierce wrote, however, the balance was shifting. The price of signal processing hardware was so reduced by microelectronics that implementations of error-control coding were becoming economical. Computer communication promised unprecedented traffic volume and demanded high speed. Take a close-to-home example: telephone lines were originally designed for human speakers, who are not only message sources but also double as encoders and decoders. When the lines are noisy, they automatically speak louder, more slowly, and more repetitively to reduce misunderstanding. Computers cannot do this, not without error-controlling modems that connect them to telephone lines. Web surfers impatient with "world wide wait" clamor for high transmission rates. Demands such as these, coupled with the availability of inexpensive hardware, have thrown information theory into high gear since the 1970s. Powerful error-control algorithms have been developed. The trellis codes introduced in 1982 and turbo codes a decade later use sophisticated mathematics to improve performance at reduced processing complexity. Implemented in modems for personal computers, they are pushing data transmission to the capacity limit of telephone lines, as Shannon predicted.[29]

Signal processing being ubiquitous in information technologies, applications of information theory spill from communication into other areas, for instance reliable computer data recording and storage. "Communication links transmit information from here to there. Computer memories transmit information from now to then," remarked communication engineer E. R. Berlekamp. Information theory provides the foundation of data compression, which saves storage space, and error-control coding, which combats the noise in writing, reading, and deteriorating storage media. Commemorating the fiftieth anniversary of information theory, Sergio Verdú observed: "With the inexorable advance of technology, Shannon's fundamental limits become increasingly relevant to the design of systems in which resources such as bandwidth, energy, time, and space are at a premium." Through patient research and development, the once arcane mathematical theory is now highly practical in both communication and computation. Gallager reminded engineers that the rapidity of today's technological development is possible only because it capitalizes on knowledge and techniques accumulated through long and patient research: "A modern cellular system, for example, is based on various concepts and algorithms that have been developed over decades, which in turn depend on research going back to 1948."[30]

Estimation in Stochastic Control

When a signal emerges from a communication channel contaminated by noise, the receiver processes it and tries to give the best *estimate* of its uncontaminated form—"best" in the sense of having the smallest probability of error. Estimation is an indispensable part of communication systems. However, it is not central to information theory that addresses systems in which all parts cooperate; the effort of signal design at the transmission end lightens the job at the receiving end. Many other situations exist in which the sources of signals are uncooperative or even adversarial, for instance in radar tracking of evasive aircraft or sonar listening for secretive submarines. In such cases estimation becomes paramount.[31]

For a general problem of estimation, consider measuring a changing system through time. The system may be deterministic or stochastic. In either case the measured results are inaccurate because the measuring equipment is subject to random noise. The task of the estimator is to obtain a good estimate for the system's state at various times based on the measured data, old and current. Sometimes the data are supplemented by other available information, such as a statistical model for the inaccuracy of measurement. The performance of an estimator is usually measured by the mean square deviation between its estimates and the system's actual state over a certain time interval. The estimator that minimizes the mean square deviation is called optimal.

Estimation is inherent in all measurements, and its problems were recognized at the beginning of modern science. Galileo conceded the inevitability of error and inaccuracy in measurements. At the beginning of the nineteenth century, Gauss introduced mathematical methods, including the least-square criterion, that determine planetary orbits from inaccurate astronomical observations. Estimation theory received another boost from the infusion of the probability calculus. Working on anti-aircraft fire during World War II, Wiener introduced in his "yellow peril" what is called the Wiener filter, which uses the whole history of data to estimate the current value of a stationary stochastic process. Many estimators are called *filters*, harking back to their root in communication, where engineers try to filter signals from noises.

As methods and algorithms for signal processing, estimators and filters are applied all over science and engineering. One area where it receives most theoretical attention is control. Knowing that they often lack complete knowledge of the systems whose behaviors they want to control, engineers developed stochastic control theory, which uses the probability calculus to reckon with uncertainty. They need to monitor the system performance in feedback control, but their sensors are often noisy and capable only of partial observations. Estimation is crucial in such circumstances. Consider the autopilot that controls an air-to-air missile. The missile has radar to measure the target position and velocity and it has a model for aircraft acceleration, but these tools are not enough. An estimator implemented in an onboard computer combines measured data, model calculation, and other available information past and current to estimate the target state. This it combines with the estimate for the state of the missile itself, and feeds the result into the autopilot that steers the missile.[32]

Theories of stochastic control with imperfect measurements attained a new height in 1960 with the introduction of the Kalman filter. It used sophisticated mathematics but was not as "perilous" to engineers as the Wiener filter, produced as it was after more than a decade of engineering science and education reform. Immediately snatched up by aerospace engineers, it made its practical debut in 1961 in the feasibility study for the Apollo space program. It was a part of the onboard computer that guided the descent of the Apollo 11 lunar module on the moon, where it helped to integrate the data from onboard sensors with information from Earth-based radar. Developed and extended by many engineers, the Kalman filter is now employed in modern control systems everywhere, from seismic research to chemical processing plants.

Its most prominent applications are in navigation and guidance of vehicles, for instance in the gyro system of the Boeing 777. If you use the Global Positioning System, chances are you have a Kalman filter in your receiver.[33]

Theories of What Can Be

Information theory provides a broad framework for communication systems, Kalman filtering a solution to a type of estimation problem. Shannon wrote his paper on "the fundamental problem of communication" and Kalman on "an important class of theoretical and practical problems in communication and control." Yet both said somewhere that they were not keen on specific applications.[34] They were engineering scientists, deep thinkers with long visions about what is generally applicable in broad types of practical situations.

Physical theories support contrary-to-fact hypotheses, but being theories of *what is,* they focus on spelling out the details of natural laws. Engineering systems theories respect the constraints of physical laws, but being theories of *what can be of use,* they abstract from most physical details to allow maximum freedom for ingenuity to create useful artifacts. *What is* is a substantive question that can be answered by physical experiments. *What can be* is a dispositional question that implies myriad possibilities, many of which are not susceptible to experimentation. That no one has yet made something does not imply that it cannot be; history has made jokes out of many hasty proclamations of impossibility. More specifically, claims of optimality are difficult to check experimentally; it is usually not feasible to build all specified options and compare their performances. All the same, the lack of experimental discipline is not a license for engineers to turn dreamers and engineering to hype. Systems theorists look into a future fogged with uncertainties, and some of their assertions cannot be checked, at least not immediately. Knowing their epistemological predicament and wanting their assertions to be nevertheless accountable, they double the effort to lay out their assumptions clearly and their reasoning precisely. For this, mathematics is the ideal language. The theorem-proof style of systems theories is not merely an academic fashion. It serves to enforce sufficient rigor of assertion and reasoning about what can be.

Calculations in physical theories tend to produce specific solutions.

Theorems in systems theories tend to assert rather general conditions, especially bounds and limits. Instead of describing this or that case, systems theories tend to give the upper and lower bounds of what is possible. A bound is derived via mathematical proof from a set of assumptions, which may include the modeling of some general physical conditions. A tighter bound may result from a different set of assumptions. Delimiting what can and cannot be achieved, these bounds are like grand strategies that point out fruitful directions for engineering research and design. An example is Shannon's channel coding theorem proving the channel capacity to be the highest data rate—the upper bound—for reliable communication over noisy channels. Pierce explained its value: "Like the laws of thermodynamics, information theory divided a world into two parts—that which was possible and that which was not. Often these were separated by a gap between upper and lower bounds, but the general geography was clear. Ingenious people no longer invented coding or modulation schemes that were analogous to perpetual motion. But, they were offered the novel possibility of efficient error-free transmission over noisy channels."[35]

Grand strategic concepts are also found in control theories. Besides optimality, there are concepts such as controllability and observability, both introduced by Kalman. A system is controllable if it can be driven by the input from any state to any other state in finite time; observable if its state at any time can be determined from the finite history of its output. These dispositional properties are difficult to verify experimentally even in existing systems, for they refer to *any* of infinitely many possible states. Fortunately they are manifest in certain characteristics of their mathematical models. By spelling them out, control theory poses the conditions for effective control, against which design engineers should check their work to avoid laboring in vain or in error. They are especially valuable for complex systems. A $10 million plasma fusion confinement experiment in Los Alamos in the mid-1960s failed because the designers of its controller did not check the dual conditions of controllability and observability.[36]

The ultimate aim of systems theories is engineering. Thus when necessary theorists sacrifice generality for practicality, settling for general conditions that are limited but sufficient for addressing useful problems. Such a grand tactical choice of conditions often enables them to go all the way from abstract mathematics to practical situations to concrete implementations. Kalman filtering is a case in point. It limits itself to linear systems. Although not all satisfactory, linearity is a good enough approximation for many important applications. Working within the theory's region of validity, engineers are able to pinpoint crucial estimation problems, bring sophisticated mathematics to bear, and develop off-the-shelf technologies for useful devices.

6.3 Wind Tunnels and Internet Simulation

Theory and experiment are often called the two legs of science. The legs are interrelated. Theoreticians keep a close eye on experimental data, calculate quantities intended for experiments, such as scattering cross sections, and use thought experiments in their arguments. Experimentalists do far more than merely test theories. They are just as creative in their new ideas, and their discoveries often surprise theoreticians. Although impatient about theoretical formalism, they nevertheless have ample mathematical ability to grasp theoretical results and perform relevant calculations for designing experiments and analyzing data. Resourceful in their own ways, experimentalists and theoreticians complement each other when they join forces to advance science and technology.

Among the natural sciences, physics is the most theoretical, but even there theoreticians constitute only a third of fresh Ph.D.'s and less than a third in the workforce. This ratio is roughly mirrored in accomplishments. Slightly less than a third of physics Nobel Prizes have been awarded for theories. The balance have been awarded for experimental and observational discoveries, new techniques, instrumentation, and practical technologies, including telegraphy, the automatic regulator, the transistor, the integrated circuit, the electronic microscope, and fast photonics. Theories have been honored as much for explaining observed phenomena as for predicting new ones; experiments as much for confirming predictions as for discovering new phenomena. It is significant that about a third of the prizes have been awarded for new measuring methods and instruments, perhaps coupled with the discoveries they made possible: the maser, the cyclotron, holography, particle detectors, bubble and cloud chambers, electron and scanning tunneling microscopes, ion and atom traps, nuclear magnetic resonance, the X-ray and electron diffraction, light or particle beam scattering, various techniques for precision measurement, methods of achieving extremely high pressures and low temperatures.[37] In these areas the thinking of experimental physicists overlaps with that of engineers. Both aim at inventing ways to achieve some purpose, especially to measure something. Although the physicists' initial aims may be purely scientific, many of their measurement mechanisms later find wide practical applications. For instance, X-ray crystallography has become a powerful tool for studying molecular structures, including the double-helix structure of DNA molecules. It and nuclear magnetic resonance are the workhorses for drug discovery in the pharmaceutical industry.

Scientific and Engineering Experiments

Experimentalists do not passively look *at* but actively look *for* something, which may be as abstract as the explanation of a puzzle. Field observations require careful selection of targets, as when astronomers decide where in the infinite universe to point their telescopes. A laboratory experiment further involves careful manipulation of instruments to create the condition that isolates a particular phenomenon whose characteristics are to be measured. An experiment is a question posed in terms of physical equipment, by which researchers wring answers from the physical world.

Good questions elicit good answers. Design of good experiments is paramount for the success of science, and good designs demand clear conceptions. Experimentalists must choose the phenomena to be investigated, the mechanisms to be used, the conditions to be set up, the quantities to be measured, the parameters to be varied, the instruments to be deployed. Besides physical insight and technological expertise, these decisions always involve background scientific knowledge and intuition. This does not imply that they are determined by a specific theory with definite predictions. A systematic theory often comes only in the wake of discoveries made by successful experiments.

As scientists mobilize increasingly complicated physical mechanisms to probe delicate phenomena and perform experiments in extreme conditions, they demand increasingly sophisticated instruments at affordable prices. This is where engineering and technology contribute vitally to the advancement of natural science. Even when a mechanism such as laser scattering is in principle capable of high resolution, it cannot be accomplished without adequate equipment to generate the required energy source, detect the weak signals, and control the apparatus precisely. Furthermore, output from sensors and detectors has to be processed and presented in intelligible form. As big sciences spawn experiments of enormous scale, automation becomes a necessity. Elementary particle physics laboratories, with their huge accelerators, are teeming with engineers. The human genome project can finish in two-thirds of the scheduled time mostly because of advanced instrumentation and automation. Control, signal processing, and automation are the forte of engineering. Makers of scientific and measurement instruments were among the forerunners of modern engineers, and now they lead a large industry with annual sales exceeding $90 billion in the United States.

Besides in engineering research, scientific experiments are also performed in the conception stage of a design process, when engineers explore possibilities for basic design concepts and improve theoretical models for particular systems, especially complicated ones. Exploratory experiments are used to determine system parameters and performance indices, demonstrate hardware, and validate system requirements. Testing and evaluation accompany all phases of development, to aid in detailed design, measure component performance, and assess system behavior. Their scope is narrower, focusing on elements in the specific system under development.[38]

Scientists and engineers usually regard experiments and tests as the highest court of appeal, but they are not unaware of epistemological problems. Experiments are mediated and susceptible to interference and error, their designs require conceptualization and their data interpretation. Knowing that measurement errors cannot be absolutely eliminated, experimentalists record imprecision by the spread of data points, accompany the points with error bars, and refrain from claims beyond the error margin. To account for random errors arising from minor interference pulling in whichever way, they develop probabilistic and statistical methods. To avoid systematic errors stemming from spurious elements built into their experimental setup, they calibrate it against established standards. If independent measurements on the equipment reveal a certain bias, they take care to subtract it from their data.

Accurate measurements of singular values, such as the charge of an electron, are harshest on equipment, but singular values are not the target of most experiments. Like mathematics, experiments emphasize *relations* among variables. Most measure how the value of a variable changes with the value of another variable, for instance the variation of pressure as a function of density. Or they measure how a relationship changes with a third variable, such as how the pressure-density variation changes as temperature rises. The variation itself can be regarded as the ratio of values with respect to a reference value, which absorbs most systematic instrumental biases, leaving the *varying trend* relatively error free. This is just one way in which experiments are designed to minimize the probability of bias on data.

A physical cause has many effects, and many causal factors occur in any situation. In most modern experiments, the measuring instrument itself involves complicated mechanisms that must interact with the target investigated. Entanglement, however, does not imply the impossibility of effective, albeit approximate, analysis. Quite the contrary, multifarious causal connections can offer the strongest support for the efficacy of experiments, if we properly distinguish factors and account for their *variations*. Good experimental designs usually succeed in suppressing major extraneous factors to isolate a range of phenomena among which the target is prominent. Then experimentalists demand that the data be *robust* in the sense of being sensitive to variations in the target parameters but insensitive to variations in other factors that are embedded in the experimental conditions.

Reproducibility, a crucial criterion for accepting experimental results, goes way beyond repeating an experiment on the same equipment. The experimental target has many other effects besides interacting with the instrument, and the instrument has many effects besides probing the target. Therefore an experimental phenomenon should be reproducible with other apparatuses, other experimental designs, or other physical mechanisms, for instance with resonance absorption and with inelastic scattering. Conversely, experimental equipment should prove its trustworthiness by working on many kinds of phenomena, as X-ray diffraction does on the structure of both cubic crystals and folded protein molecules. A diversity of approaches provides greater support, because they complement each other's shortfalls; by comparing variations, experimentalists can spot and correct spurious peculiarities in their own setups. It is like triangulation; to determine the distance of an object, we measure from two or more connected viewpoints, and the accuracy of measurement increases with the length of the base of triangulation. This is the experimental equivalent of thinking about one phenomenon in terms of many theoretical representations. Extensive cross-checking from many approaches and many researchers cannot produce absolute certainty, but it can reduce the probability of error to minuscule levels.

Wind Tunnel Experiments for Transonic Flight

An engineering experimental method with wide applications is the wind tunnel. Although everyone has experienced breezes and gusts, for a long time people had only crude knowledge of the interaction between air flow and solid bodies. Equations for fluid dynamics, produced in the mid-nineteenth century, are nonlinear and defy analytic solutions. Novel concepts such as Prandtl's boundary layer, which identify physically significant approximations, marked major steps in theoretical aerodynamics. However, even with today's supercomputers, practical solutions in the presence of turbulence fall short. Experiment remains firmly with analysis and computation as one of three pillars of aerodynamics.[39]

Many kinds of aerodynamic experiments exist, from swirl arms and drop bodies to actual flight. The laboratory choice is the wind tunnel, in which an air current with controlled characteristics flows past a model sitting in the test section, called the tunnel's throat, which is equipped with sensors to measure various forces arising from the interaction between the air and the target. Besides the aerodynamics of aircraft and cars, the wind tunnel is used to study wind power devices, dynamic loads on buildings, dispersal of pollutants, and countless other phenomena. Its basic principle was enunciated by Leonardo, who thought deeply about flight: "The action of the medium upon the body is the same whether the body moves in a quiescent medium, or whether the particles of the medium impinges with the same velocity upon the quiescent body."[40] The principle waited three centuries for the rise of aeronautics. The first wind tunnel appeared in 1870 and wind tunnel experiments contributed crucially to the Wright Flyer of 1903. Impressed by Wilbur Wright's public demonstration, Eiffel built a wind tunnel in the shadow of his tower. Among other pioneering measurements, he confirmed Leonardo's principle by comparing wind tunnel data with behaviors of bodies dropped from the Eiffel Tower, and by comparing data for complete aircraft models with performances of real aircraft in flight. The diversity confirmed the effectiveness of the wind tunnel for detailed measurements.

Conducting a wind tunnel experiment is far more complex than blowing a huge fan at things. To start with, experiments are mostly performed with small models. Scaling the results up for full-size objects is addressed mostly by dimensional analysis. Instead of calling for fancy mathematics to solve the fluid equations, dimensional analysis depends on deep physical insight into the relations among dynamic variables and elementary mathematics to exploit similarity of dynamic flows. Such theoretical acumen is fruitful not only in experiments but in science generally. It establishes clear connections among various experimental conditions so that their results can be collaborated. Furthermore, it sweeps theoretical difficulties under parameters such as lift and drag coefficients. By measuring the values of these parameters, experiments cut the Gordian knot to the theoretically intractable problem.

By the 1930s, the wind tunnel was a maturing technology with a remarkable track record in aerodynamics research and aircraft design. Then new problems appeared. To simulate realistic aeronautical situations, a wind tunnel had to provide a uniform air flow over the model. This was readily achieved for flow speeds far below Mach 1, the speed of sound, where the air can be approximated as an incompressible fluid. It is different for flow speeds in the transonic region roughly between Mach 0.7 and 1.3, where shock waves appear. Data show that the lift coefficient drops and the drag coefficient rises abruptly. A chart similar to the one in Fig. 6.3a, when shown to the press, led to the misleading notion of a "sound barrier."

The sound barrier claimed its first victim in 1941, when a test pilot pushed his plane beyond Mach 1 in an operational dive and crashed. Other accidents followed. The need to understand transonic aerody-

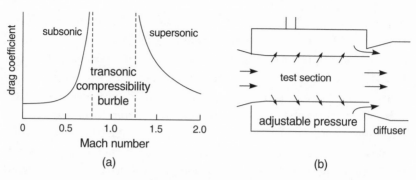

Figure 6.3   (a) The drag coefficient rising steeply in the transonic region spoiled traditional wind tunnel experiments. (b) The problem is solved by venting the test section of a transonic wind tunnel with carefully designed slots or holes. *Source:* (a) Adapted from J. D. Anderson, *Introduction to Flight,* 4th ed. (Boston: McGraw-Hill, 2000), p. 406; (b) adapted from A. Pope and K. L. Goin, *High-Speed Wind Tunnel Testing* (New York: Wiley, 1965), p. 106.

namics became more urgent as the advent of turbojet and rocket propulsion promised high-speed flight, but investigation was stone-walled. Theoretically, the fluid equations for compressible fluids were recalcitrant until the 1970s, when they were cracked by computers. Experimentally, wind tunnels were choked. Shock waves generated at the model reflected off the tunnel walls and interrupted the mass of air flow through the throat, ruining experiments. Removing the walls did not help; shock waves converted into expansion waves at the free-air boundary and returned to strike the model, making the data worthless. The wind tunnel as an experimental instrument was compromised by the very phenomenon it was used to study. Researchers were frustrated by the difficulty but not deceived. They could not explain what happened, but they clearly identified the trouble as spurious behaviors of the wind tunnel.

The lack of understanding hampered adequate modifications of the wind tunnel, and the result was a transonic experimental gap. Data were unavailable where they were most needed. Aeronautical engineers refused to accept defeat, however. The National Advisory Commission on Aeronautics' (NACA's) Langley Aeronautical Laboratory was a leader in wind tunnel research, and always insisted on using all available techniques to gain understanding. Desperate for knowledge,

Langley engineers resorted to various stopgap methods. John Stack, head of high-speed wind tunnels, proposed a research aircraft that would gather the coveted aerodynamics data in flight. He persuaded the Air Force, which was enticed by the research tool's doubling as a demonstrator for transonic flight. An exemplary collaboration among researchers, the military, and industry produced the Bell Aircraft X-1 that, piloted by Chuck Yeager, broke the sound barrier in 1947.

Participating in the design and operation of the X-1 and quickened by it, Langley engineers pressed harder on wind tunnel research. By reducing model sizes, inventing new support systems, and designing specially shaped throats, they steadily narrowed the transonic experimental gap. Yet a gap still remained between Mach 0.95 and 1.1. The final assault was headed by Ray Wright. Steeped in Langley's collective experience and combining it with his own theoretical intuition, he argued that the chocking problem could be solved by introducing ventilating slots in the test chamber of the wind tunnel, as illustrated in Fig. 6.3b. Although his idea met with considerable skepticism, he had the full support of Stack, who fought for converting two large tunnels to slotted throats. To design exact slot configurations for smooth transonic flow distributions took years and a patient mixture of testing, calculation, and art. By 1950 the Langley team had closed the gap and turned the wind tunnel into a successful instrument for all flow speeds from subsonic to supersonic.

Laboratory data on transonic aerodynamics came just in time. It was one thing for the X-1, a light air-launched aircraft with rocket propulsion, to overcome the transonic drag hurdle. It is another for ground-launched fighters with turbojet propulsion to do so with fuel left for supersonic operation. Intensive experiments in Langley's new slotted-throat wind tunnels helped to cut transonic drag by up to 75 percent, which was crucial for the success of the first supersonic jet fighter.

The struggle with the wind tunnel shows that even when knowledge is grossly incomplete, experimentalists are able to tease apart instrumental malfunctions and anomalies of the target phenomenon, based on rudimentary theories and physical intuition gained from a variety of experiences. As obstacles fall one by one, both knowledge and instrumentation improve, leading to better experimental designs and more reliable data. Ramifications of engineering experiments spread through a great variety of practical systems, which in turn provide even more feedback and confirmation.

Modeling, Simulation, and Computer Experiments

Relations between theories and experiments are tightened in the new research avenues opened by computers, so much so that in some areas it is difficult to tell whether a project is theoretical or experimental. A case in point is computer simulation, which includes both *representing* a real situation by a model and *implementing* the model to run on a computer.[41]

Simulations can be theoretically or experimentally oriented. In the former, researchers use computers extensively to solve equations in well-established theories such as aerodynamics or electromagnetism. Designed to tackle complex cases that resist analysis, these computer models invariably depend on some physical and numerical approximations, such as the finite-element method used to solve Maxwell's equations for complicated device configurations.

Many simulations not so heavily based on theoretical equations are aptly regarded as virtual experiments. Like physical experiments, they obtain output of particular instances under specific conditions, except they are abstract and exist only in cyberspace. They constitute a new technique, which is most useful for studying complex dynamic systems with large numbers of interacting components. Consider, for example, the traffic on a highway system, with cars crossing each other's paths on their way to disparate destinations. Researchers are interested in overall traffic patterns, but they can conceptually describe only the motions of cars, individually or in interaction with a few others. Even these descriptions are often approximate and statistical, because the behaviors of drivers are not fully known. Based on general knowledge and detailed data gathered for the system under study, researchers frame a conceptual or mathematical model that captures essential features on the car level, either deterministically or probabilistically with the help of random numbers. Then the model is translated into a simulator, which is in a form that can be executed on a computer. The computer tracks the behaviors of thousands of cars, and thus simulates or imitates real traffic.

A simulator is a virtual laboratory. Equipped with one, researchers design an experiment by specifying a set of parameters, such as road condition and traffic load. To run the experiment, they set the cars in motion and observe what happens on the highway system. They vary the parameters, perform many experimental runs, and statistically extract the overall patterns of the system. This is a kind of experimental exploration. Its mathematical model reflects the physical idea that the system is made up of components, but involves minimal theoretical assumptions on the system level. Nevertheless, the computed results reveal information about the system as a whole by making numerous observations on the components under controlled conditions. Computer simulations are used to study and evaluate a wide range of topics: air traffic and other transportation systems, logistic and other military systems, supply chains and other manufacturing systems, integrated circuits and other computer systems, radio networks and other communication systems.

An important area in engineering research is the operation of computer networks and the evolving Internet. Empirical measurements of Internet traffic reveal route instabilities and other intricate dynamic behaviors. Understanding its dynamics is crucial to designing new protocols and controlling operations, to prevent dreaded gridlock and optimize network performance. For these tasks network engineers increasingly rely on network simulation.

As discussed in section 3.3, the Internet uses packet switching, which breaks a message into small packets and sends them via whatever routes available. Most network simulators focus on the level of packets, tracking packets as they emerge from their hosts, move via routers to their destinations in the network, and exchange information with various stations. How they move and exchange information depends on the characteristics of the network topology, links, and protocols, specifications of which would define various experiments. Some network simulators can accommodate tens of thousands of hosts and routers grouped into interlinked local area networks of various configurations. Although minuscule compared with the real Internet, which had about 100 million hosts in 2000, they are large and heterogeneous enough to reveal many interesting behaviors. Simulators also provide tools for abstraction and scenario generation, so that network engineers can design experiments to target a particular level of network operation. An experiment to evaluate the performance of internet protocols would feature realistic routers between networks but abstract from some details of local area networks and most details of the physical link. An experiment to evaluate the error and delay rates in a wireless network would focus more on the interaction between the characteristics of the physical link and the transport control protocol. It is akin to researchers in a physical laboratory rearranging their equipment for different experiments that measure different quantities.

Despite their abstractness, simulators and the experiments they support address real-world problems. To ensure that they do so is a most important task that is common to most software development. That task has two parts, called validation and verification, to address the two steps in a simulating process, modeling and computer implementation. *Verification,* which eliminates programming errors, ensures that the computer code faithfully implements concepts in the model, that it is doing things right. *Validation,* which evaluates how accurately the model reflects the real-world phenomenon it purports to represent, ensures that the right thing is done. Sometimes models can be validated by comparing specific simulation results to real-world experiments, but this may not be feasible for simulations of large systems such as the Internet. Engineers and scientists have to devise various methods to validate models. Furthermore, they have to find criteria for deciding how good is good enough for models for various types of problems, because models are approximate and higher degrees of accuracy come with higher costs.

6.4 Integrative Materials Engineering

Geologist Scott Montgomery wrote: "[Geology] is unique in defining a realm all of its own yet drawing within its borders the knowledge and discourse of so many other fields—physics, chemistry, botany, zoology, astronomy, various types of engineering and more (geologists are at once true 'experts' and hopeless 'generalists')." Quoting the remark, material scientist Robert Cahn observed that the description applies to his discipline as well.[42] Basic academic disciplines such as physics and biology are drawn according to the most general distinctions in phenomena. Concrete complex phenomena, natural or artificial, usually involve mechanisms studied by several basic disciplines. To bring knowledge from several disciplines to bear on a single realm is characteristic of many sciences, natural and engineering.

Consider two fields: stellar physics and nuclear engineering. To study the structure and evolution of stars, astrophysicists analyze stellar behaviors into fusion and other nuclear interactions, energy and matter transport, gravity, and other relevant mechanisms ranging from the microscopic to the macroscopic. Adding constraints peculiar to stellar configurations, they synthesize relevant principles to explain the behaviors of various types of stars: hydrogen-burning stars in the main sequence, red giants, white dwarfs, supernovae, neutron stars, and so on. Nuclear engineers also consider natural forces. They analyze reactor principles according to fission and other nuclear interactions, energy and matter transport, structural containment, and other relevant physical mechanisms. Then they consider constraints imposed by reactor configurations and practical considerations such as safety, fuel availability, waste disposal, and economic competitiveness. They synthesize the physical principles and practical constraints in exploring various reactor types: light-water, heavy-water, breeder, pebble-bed, and so on, some more suitable for generating electricity, others for powering ships. Obviously studying nuclear fusion reactors in the sky is different from designing fission reactors on the ground. Nevertheless, on a general level stellar physics and nuclear engineering reveal certain similarities in the ways they integrate various factors and branches of knowledge to form a science specialized to a distinct realm.

Scientific synthesis of fundamental principles to explain or predict complex phenomena is a creative activity. One must use novel concepts to cut through the maze of complexity and grasp important factors. Geological concepts such as plate tectonics and astrophysical concepts such as supernovaic explosion are not found in basic physics. They are the substantive content of particular sciences. The same holds for engineering sciences. Stellar physicists and nuclear engineers separately develop the substantive details of nuclear physics for their respective topics.

A lost hiker is relieved to see a pile of stones. Stone piles are everywhere in nature, but artificial piles have certain readily recognizable peculiarities. Artificialities useful for productive purposes are systematically represented in the engineering sciences in addition to natural laws. Vincenti provided an example. Steam engines, gas turbines, and chemical reactors all involve simultaneous thermal and flow effects in their essential regions. Many such regions have fins, nozzles, and other knotty configurations that frustrate boundary-value solutions of the flow. To address them, engineering thermodynamics introduces the concept of *control volume* enclosed in an imaginary fixed boundary. The concept, which enables engineers to ignore the complexity of internal flow and calculate overall effects important to engineering design, is absent in physics. In short, although the term "thermodynamics" is common to physics, as an engineering science it contains many novel concepts of its own.[43]

Structure, Property, Process, and Performance

For lack of materials capable of withstanding high stress and temperatures, Frank Whittle's proposal for an aircraft jet engine was rejected by the British Air Ministry in 1929. Optical communication had to wait for low-loss glass fiber before it could take off in the 1970s. Holographic storage of hundreds of gigabytes of data in a cubic centimeter, invented in 1994, is still waiting for an appropriate material for commercialization. Advanced materials are always critical to advanced technology, no matter where the frontier lies. Materials constituted one of six engineering sciences identified in a 1955 ASEE report. Since then materials science and engineering (MSE) has become a multidisciplinary department in many universities.[44]

The whole universe is matter and its equivalent, energy. Material is not merely matter, as metallurgist Cyril Smith argued. Matter is substance; it is what is. Material further connotes function and utility. It is matter with certain properties that make it useful for certain devices and structures.[45] While philosophers ponder substance in contrast to form, physicists pursue the basic building blocks of matter, and chemists study the transformation of one substance to another, materials fall more into the arena of engineers. Of course materials are inseparable from matter. Development of advanced materials is impossible without scientific knowledge about matter. This was emphasized in a 1989 study by the U.S. National Research Council (NRC): "Science and engineering are inextricably interwoven in the field of materials science and engineering."[46]

Materials come in all varieties and can be classified in many ways. According to their nature, they fall roughly into four big classes: metals and their alloys; ceramics, such as pottery and cement; polymers, such as wood and polyethylene; and composites, such as paper and fiberglass. They are also grouped by their applications. Structural materials for buildings and vehicles are valued for their mechanical properties, such as strength and elasticity. Functional materials are valued for their electrical, electronic, magnetic, optical, and other properties. Semiconductors form a distinctive class because of their importance in information technology. MSE studies not only bulk materials but also surfaces, interfaces, and thin films such as those used in many microelectronic devices.

The first materials people used occurred naturally. As civilization advanced, people found ways to extract more materials from nature and process them for better performance. Material processing techniques were so important they marked major historical epochs: the stone age, bronze age, iron age. Museums brim with artifacts expressing the marvelous skill with which practical artists processed metals and ceramics to achieve superb beauty and function. Complicated metallurgical processes, perfected in the thirteenth century, produced the Japanese samurai sword, whose performance was unmatched until the twentieth century brought scientific metallurgy. The vulcanization of rubber, electrolytic isolation of aluminum, synthesis of plastics, and countless other processes that produce materials with desired properties constitute a vital component in technological progress.

To introduce concepts and measuring methods for material properties was an early major task when modern engineering emerged in the seventeenth century. The job was arduous. Many properties of matter useful for materials are either not obvious or difficult to systematize. It was not until around 1800, when Thomas Young introduced the concept of elastic modulus, that a crucial property of structural materials was quantitatively captured. Along with quantitative concepts, techniques and instruments were developed to measure subtle properties accurately.

Besides systematic studies of material properties and processes, an extra ingredient is required for modern MSE. The crucial element, as Smith remarked, is to understand how a material's *macroscopic properties and processing techniques* connect to its *microscopic structures and mechanisms*. The connective idea surfaced in the seventeenth century but had no influence when atomism itself was a speculative philosophy. Then science marched into the molecular realm with chemistry in the lead, followed by quantum mechanics and solid-state physics. The crystalline structures of many solids, mathematically described and classified in the nineteenth century, were confirmed by X-ray diffraction in 1912. Theories of metals and crystal lattices appeared. Chemistry unraveled the macromolecular structures and reactions of polymers in the mid-1930s, when nylon was synthesized. Metallurgy turned from a craft to a science in the 1920s, stimulated by technological demands on metals and a flourishing steel industry. It absorbed principles of physical chemistry and phase equilibrium, developed theories of alloys, and led the way in relating metallurgical processes and macroscopic performances of metals to their microstructures. The first electron microscope, in 1931, followed by many other inventions, enabled researchers to probe deeply into microstructures. For decades after the 1940s, atomic energy spurred a flurry of R&D on radiation damage to materials. Invention of the transistor in 1947 ignited a even bigger effort that combined solid-state physics with device performance and material processing. Armed with computing power, scientists began to tackle explanatory models of increasing complexity.

Materials-related studies share certain core ideas, if only because all materials are made up of atoms and atomic bonds. However, research activities were scattered in various academic departments. Combination and integration began in 1959 when Northwestern University established the first department of materials science, with engineering added soon afterward. From the beginning, MSE departments have been multidisciplinary, drawing knowledge from chemistry, physics, metallurgy and other branches of engineering, and, increasingly, medicine and biology. More important, they distill and synthesize relevant knowledge from diverse disciplines to create a coherent thrust toward advancing both the understanding and the design of new materials for new technologies.

The integrative nature of MSE was apparent in the 1989 NRC report, which defined it as a synergy of four elements: "It is these elements—properties, structure and composition, synthesis and processing, and performance and the strong interrelationship among them—that define the field of materials science and engineering."[47]

MSE undertakes basic research to attain fundamental scientific knowledge, but not purely for its own sake. Its utilitarian orientation is underscored by its attention to performance and processing. Demand pull and scientific push are usually cited as distinct forces that drive technology. They are integrated in MSE, which is like a heavy train with two engines, one pulling in front and the other pushing at the back, combining their forces to climb a steep grade.

An example of the duality of application and basic science in MSE is high-temperature superconductivity. Experimentally discovered in 1986, it excited both fronts. Intrigued by the surprising phenomenon, theoretical physicists flocked to explain its underlying mechanisms, which have so far defied their efforts. Enticed by its utilitarian potential, industries and governments competed to invest in R&D. Materials with zero electrical resistance at practically achievable temperatures can boost the efficiencies if not revolutionize the designs of electrical and electronic devices. In the most obvious case, superconducting power lines can eliminate transmission loss, which now dissipates about 7 percent of all electricity generated. Unfortunately, high-temperature superconductors, which are ceramic, are fragile and difficult to process. Despite heavy government subsidies, not until 2001 were superconducting cables connected to the utility grid, first in Copenhagen, then Detroit and Tokyo. At roughly fifty times the cost of copper cables, however, their use is confined to short distances in crowded cities, where tearing up streets to add transmission capacity is even more expensive. One reason for the high cost is the lack of scientific understanding; developers are forced to proceed by trial and error, which misses many possibilities. High-temperature superconductivity exemplifies the intricacy of materials and the need for basic research for their efficient use.[48]

Mechanisms in Many Scales

To see MSE as an engineering *science*, consider two of its four elements: structures and processing. *Structures* encompass characteristics static and dynamic, equilibrium and nonequilibrium. A material's structure includes the types of atoms it contains and the arrangements of atoms that underlie its potentially applicable properties. It also includes the forces and mechanisms that arrange atoms regardless of their potential for applications. Research into material structures discovers basic scientific principles and hence overlaps with various natural sciences.[49]

Processing encompasses the making and manipulation of materials. It includes assembling atoms and molecules into a material, in which case it is also known as *synthesis,* especially when chemical means dominate the process. It also includes processes that form materials on more macroscopic scales, such as sintering and joining. The planar process so vital to the success of microelectronics depends on many semiconductor processing techniques and knowledge of their underlying mechanisms.

Structures and processing affect each other. How a material is processed strongly affects its resultant structures, which determine its properties and performances. Conversely, scientific knowledge about structural mechanisms facilitates the development of appropriate processes to make materials with desired performances.

Structures in materials span a hierarchy of levels from the electronic and atomic to the macroscopic. Mechanisms occur at various levels, from quantum mechanics, chemical reactions, and kinetics to thermodynamics and continuum mechanics. They proceed at different temporal scales and generate patterns on different spatial scales. Yet they are connected to each other. Framing theoretical models that take into account multiple levels and their interrelations is at the frontier of MSE research.

Consider an old problem that motivated Galileo's *Two New Sciences:* things break. Material fractures have caused building collapses, airplane crashes, and other disasters. Breakages and the constant maintenance and repairing required to prevent them cost about 4 percent of America's gross domestic product, according to a 1983 study by the Commerce Department. How things break is a tough problem that

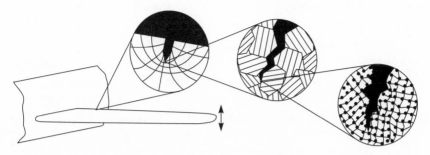

Figure 6.4 A crack in an aircraft's wing studied on many levels.

must be tackled on many scales, as illustrated in Fig. 6.4. On the macroscopic level, elasticity and other theories describe the load and its dynamics, for instance how the vibration of an airplane's wing affects its material. At the vicinity of a tiny notch, continuum mechanics describes the uneven distribution of stress generated by the crack and the pulling apart. On a smaller scale, we see the crack running along boundaries between grains, crystallites that are often called microstructures of the material. Magnifying the tip of the crack, we see how individual atoms move to release the stress. Atoms in brittle materials snap apart, so the crack propagates as though a knife is cutting through the bulk of the material. Thus a glass cutter has only to scratch a shallow line to break a piece of glass cleanly. Atoms in ductile materials such as metals slide over each other, stopping the crack at the price of slightly deforming the material. Atomic motions are crucial to the dynamics of cracks but do not determine them completely. Also crucial are structures at larger scales such as defects and dislocations. Complete explanations are still lacking, but we do have significant scientific knowledge of the mechanisms underlying fractures. With that knowledge engineers are able to introduce reinforcing fibers, microcracks, and other remedial structures to enhance crack resistance, a most important quality in structural materials.[50]

Most engineering materials pass through a molten state during their processing. How a material solidifies from the melt critically affects its microstructures and properties. Controlling conditions for high material performance requires an understanding of the mechanisms underlying solidification. Materials engineers are sometimes successful, for instance in growing single-crystal semiconductors of increasingly large sizes and fewer defects for microelectronics. Other areas are more difficult. Solidification is a complex process that depends on a slew of subtle nonequilibrium forces that generate intricate patterns. When metal alloys solidify, they form submillimeter tree-like patterns whose characteristics are strongly correlated with alloy performance. Pattern-forming mechanisms are so complicated their representations invoke turbulence, chaos, and other cutting-edge mathematics. Theoretical research aids in the development of processing techniques such as directional solidification, which produces advanced alloys for aerospace and other applications. Furthermore, many theoretical concepts and mathematics are also used by other scientists, for example by cosmologists striving to understand the formation of patterns in the primordial universe. Materials scientist James Langer remarked: "I also find it sobering to realize that the cosmologists who are working at the furthest frontiers of natural philosophy have so much in common with the engineers who are trying to improve the manufacture of engine blocks or brake drums."[51]

Materials Engineering and Systems Engineering

To see MSE as an *engineering* science, notice that through performance it overlaps with engineering materials and systems engineering. Evaluation of a material covers its behaviors from cradle to grave. Besides reliability in active service under often harsh conditions, materials are evaluated according to how cost effectively they can be processed into useful forms and how adequately they can be recycled or disposed of at the end of service. Requirements of life-cycle performance are steadily tightening the relation between materials engineers and product design engineers.

When products are simple and undemanding, product engineers can proceed without talking to materials engineers. They can select from huge databases of available engineering materials, or specify their requirement and throw it over to the materials department. When products get complicated and when engineers push for optimality, they increasingly include material properties and processing as integral to their design considerations. This is another example of the comprehensive view of systems engineering. The optimal performance of a material in a complex system has wide ramifications. For instance, in selecting the material for a car's hood, automobile engineers consider not only structural properties such as stiffness and corrosion resistance but also electrical properties. The hood has to minimize the electrical interference between the power train and the radio antenna. If its material fails to suppress interference, engineers may have to redesign the antenna. This simple case shows how intricately materials' properties interact in complex systems. As such situations are increasingly the rule rather than the exception, product design and material design increasingly overlap.

6.5 Biological Engineering Frontiers

The spectacular ascendancy of life science strikes anyone who glances at the distribution of federal research funding among various branches of science and engineering (see Appendix B). Molecular biology is rich both in Nobel Prizes and in patents. Results of its basic research are eminently applicable to humans, who are biological beings. The new biotechnology based on it already has a glittering success record, and it is only at its beginning. Probably its greatest promises lie in health care, which is challenging, humanitarian, and lucrative. Universities woo pharmaceutical companies to fund R&D centers. Wall Street is exuberant and so are the news media, in which advances in genetics and medicine have almost monopolized televised reports on scientific discoveries. The excitement in life science and biotechnology has also influenced engineering. Bioengineering is poised to become a new engineering science.

Technologies for Life and Using Life

Biotechnology involves much more than genes, just as information technology involves much more than computers. Science and engineering brought to bear directly on living processes can involve either using mostly nonliving means *for the sake of life* or *using life as a means* to make desired products. The various meanings of the term "biotechnology" reveal engineering's diverse and deep relations with life and biological processes.[52]

Technology working directly *for* life implies developing mechanisms, living or nonliving, to interact with a living process and thereby serve it. To succeed, it requires considerable knowledge about the living process to be served. Biotechnology as man-machine interaction was introduced in 1947 as an engineering program at the University of California at Los Angeles. Extending this sense, it includes *biomaterials research* and *biomedical engineering*. Biomaterials used in everything from pill coatings to artificial heart valves may be inorganic, but they must interact satisfactorily with the biological processes in the living body to perform their desired functions without undesirable side effects. Biomedical engineers, who design instruments for medical imaging, monitoring, diagnosis, and therapy, must be familiar with the relevant physiology. At present, biomaterials and biomedical instrumentation are seldom counted as biotechnology, but the lines are beginning to blur as people increasingly use life for life.

Biotechnology now usually refers to the science-based *use* of living organisms and processes in industrial production. Crudely, it consists of three main areas, distinguished by the kind of organism used and the level at which living processes are manipulated. Manipulations on the levels of animals and plants, cells, and molecules trace their roots, respectively, to agriculture, fermentation, and genetics.

Agriculture is vital to our livelihood, despite the rise of manufacturing. Elevating the yields of agriculture and husbandry has long been a goal of science and engineering. In fact the term biotechnology was coined in 1917 by agricultural engineer Karl Ereky in an article titled "Biotechnology of Meat, Fat, and Milk Production in Large-Scale Agricultural Industry."[53] Here the focus is on technological manipulation at the level of animals and plants to improve their productivity. This kind of biotechnology is pursued in *agricultural and biological engineering*. Besides other contributions, it is essential to the practical success of genetically modified crops and animals, either for consumption or as "living factories" for producing desired drugs and other products.

The phenomenon of fermentation was used for brewing and baking in many ancient cultures. Not until 1856, however, did Louis Pasteur firmly link fermentation with physiological processes in microbial organisms. In 1897 Eduard and Hans Buchner inaugurated biochemistry by demonstrating that a yeast extract acted as a catalyst that promoted chemical reactions characteristic of fermentation in vitro, or outside of a living organism. Thereafter science was applied to transform the craft of fermentation into a technology that uses bacteria, fungi, and other microorganisms for industrial production. A big step forward occurred in the R&D project to produce penicillin during World War II. For each aspect of the project Merck, a participating pharmaceutical firm, assigned a microbiologist and a chemical engineer to work together hand in glove. For its innovations it received a prize for which the term *biochemical engineering* was first used. To synthesize new knowledge, Merck collaborated with Columbia University and Princeton after the war in developing the field of biochemical engineering, which quickly distinguished itself in the production of antibiotics, steroids, vitamins, enzymes, and other biological products. In 1961 the major journal of biochemical engineering got its present title, *Biotechnology and Bioengineering*. Biochemical engineering soon expanded into what is called *bioprocess engineering*. Focusing mainly on the cellular level, it uses living cells as tiny factories. Engineers design bioreactors to control the metabolic and reproductive functions of cells to produce biological molecules and materials.

The biotechnology that captures headlines and eclipses all others is the technology based on molecular biology of genes and cells. Molecular biology revolutionized biology by fusing two fields: biochemistry and genetics, which was developed by Gregor Mendel in 1861. Taking a chemical view of the cell and its components, including proteins and genes, molecular biologists analyze the molecular structures and mechanisms underlying their functions. In 1953, James Watson and Francis Crick discovered the double-helix structure of deoxyribonucleic acid (DNA) that embodies the genetic codes for proteins. To study and manipulate genes in the laboratory, various techniques were developed to splice a long DNA molecule at various points, make copies of a particular segment, modify a segment, combine selected segments, and insert the product into a living cell. These techniques are summarily known as *recombinant DNA technology,* or *genetic engineering.* It produced in 1973 the first genetically modified organism, a bacterium, and in 1982 its first therapeutic protein, recombinant human insulin.

Recombinant DNA technology has become a definitive element of the "biotech industry."[54] It brings new and coveted products, such as protein drugs and other biopharmaceuticals. As celebrated as it is, however, the "new" biotechnology cannot stand alone. The biotech industry depends on "old" biotechnologies to manufacture the products. Biopharmaceuticals are so sensitive to their production processes that regulators in Europe and the United States demand that they be produced for the market in the same facility that produced them for the final round of clinical trials. The extreme purity and consistency required of biopharmaceuticals would not be possible without advances in bioprocess engineering.[55]

Biologists and biochemists are the protagonists in molecular biology and recombinant DNA technology, but engineers also play significant roles. The automation of gene sequencing is an engineering feat. Now a robot can sequence more than 330,000 bases per day, a job that took 50,000 manual workers a decade ago.[56] Finding the valuable pieces of information in the torrent of results from high-throughput data factories is like panning for gold in a river. It depends so much on powerful computers and algorithms that it attracts traditional information technology firms. IBM initiated a large project to develop Blue Gene, a supercomputer designed for computational biology.[57] But let us put these rather muscular contributions aside and examine how emerging engineering sciences change not only the biotech industry but biology itself.

Bioengineering: A New Integrative Engineering Science

In June 2000, President Bill Clinton and Prime Minister Tony Blair jointly announced the completion of the first draft of the human genome, or the set of all human genes. The success of genomics marks a second turning point in biology. First biology turned from phenomenology to analysis, now it is turning from analysis to synthesis, not unlike turning the corner of the systems-engineering V in Fig. 5.5.[58]

Nineteenth-century biology was mainly description and classification. Molecular biology analyzes compositional structures and underlying mechanisms. Its methodology is often called reductionism. Chemical analysis is called reduction, and reductionism as analysis is a sound and successful scientific approach. It should not be confused with reductionism as the philosophical dogma asserting that knowledge of a system is exhausted by knowledge of its components because a system is *nothing but* the sum of its components, as we are nothing but the aggregate of our genes. This philosophical view has been refuted many times. Analysis has been carried to the bottom of biology. The human genome project provides a detailed map for the positions of some 30,000 protein-coding human genes and their exact structures as sequences of four building blocks. It advances our biological knowledge tremendously, but biologists realize that it misses many important reactions. Genes and proteins do not act in isolation. The living cell is a complex environment in which genes are turned on by external stimuli and the proteins they form undergo additional transformations. The vital dynamics of cells and organisms as wholes are not found in the "parts catalogs" detailing genes and proteins. Having decomposed the organisms into basic biological building blocks, biology makes a second turn, this time from reduction to integration. Functional genomics tries to relate the patterns of a genome to the growth of the organism.[59]

For the integrative trend in the postgenomic era, the notion of *systems biology* is being revived. Introducing a special section on the topic, the editors of *Science* explained that systems biology was an old idea dating back at least to the 1960s in the work of biologist Ludwig von Bertalanffy. It lay dormant for decades because knowledge about the components of organisms as biological systems was woefully inadequate.[60]

While systems biology slept, systems engineering galloped ahead. In designing systems of increasing complexity, engineers have become experts in relating the functions of large systems and the structures of their components, providing not only qualitative descriptions but also quantitative predictions. So far most engineering systems have been inorganic. However, because systems theories abstract from physical matters and distill general principles, they can be adapted and extended to cover biological systems as well. To achieve this, engineers must master biology, but absorbing new knowledge and putting it to work has always been their strength. Seeing the opportunities opened up by genomics, engineers are moving in to capitalize on their comparative advantage. Chemical engineer Gregory Stephanopoulos observed that "*pattern discovery* tools from metabolic engineering will have a very significant impact in structuring the intellectual content of the emerging field of *Systems Biology*."[61]

A new bioengineering that assimilates molecular biology attracts biomedical, materials, mechanical, electrical, computer, and other engineers. Above all, it attracts chemical engineers, who have been designing bioprocesses for decades. Almost a century ago, chemical engineers established their discipline by synthesizing chemistry and physics. Now they are ready to pull in molecular biology for a new synthesis. In a survey, they voted biotechnology to be the most significant advance that their discipline will make in the new century, pharmaceuticals being the second.

The new bioengineering is integrative. This does not mean bioengineers have the hubris to believe they can grasp everything all at once. Proceeding scientifically, bioengineering embraces the dual analytic-synthetic approach. Its research spotlight still focuses on a single level, be it the molecular, cellular, or organic. However, the focus is not so tight as to black out the rest of the stage. The focal level is studied as situated within the whole organism, which also contains other levels.

Biomolecular, Metabolic, and Tissue Engineering

The spotlight on the molecular level illuminates *protein engineering* and more generally *biomolecular engineering*. Proteins as enzymes act as catalysts for biochemical reactions. Designing catalysts is a staple job of chemical engineers, who naturally move to design novel proteins. Protein engineers use recombinant DNA technology and computer models of protein structures. Equally important, they consider the protein not only in the specific reaction that it catalyzes but also in the context of the entire industrial process in which the reaction plays a part. For this they integrate the protein's *structure* with its catalytic *function* in a biochemical reaction, and its *performance* in a biological environment. It is akin to the structure-property-performance synergy of materials science and engineering.[62]

Biochemical engineers have long used cells as tiny living factories. In the past, insufficient knowledge about the internal mechanisms of cells kept them out of the factories and confined them to controlling external parameters that manipulate cells as wholes. Now that the factory door has been unlocked by molecular cell biology, they can attempt to modify the assembly lines inside. Using recombinant DNA technology, they introduce new types of cells designed to perform specific industrial functions, for instance producing a protein or an antibiotic in large amounts. To achieve this they must identify the genes that code for the desired functions under certain cellular and environmental conditions. The conditions are important, for the cell is a most complex factory with many interacting metabolic pathways, and it is sensitive to many external stimuli. Just as the productivity of a stamping press can be killed by an erratic flow of raw materials in an ill-managed factory, the productivity of a gene can be killed by interfering regulations of metabolic pathways in an ill-designed cell or an inappropriate bioreactor. To enhance productivity, engineers consider not individual genes but the cellular system in its entirety. The result is *metabolic engineering;* metabolism includes all cellular activities. In the tradition of chemical engineering, metabolic engineers take the scientific route. Not satisfied with merely designing cells to optimize this or that particular production process, they aim to discover and systematize the general principles underlying the dynamic relations among cell structures, functions, and performances. The generality of the principles facilitates their judicious application to both industrial processes and biological processes in vivo, or within living organisms. Besides commercial ends, such as drug discovery, they also contribute to basic scientific understanding of metabolism, including metabolic diseases such as diabetes.[63]

Shifting the spotlight up from cells, we come to tissues and organs. Tissue loss and organ failure are major health problems that consume almost one half of health care costs in America. Surgeons and clinicians have developed remarkable ways to replace damaged tissues and organs, by implanting artificial devices or by transplanting parts from donors or the patients themselves. These operations save many lives but still leave much to be desired. Materials used in artificial devices are mostly off-the-shelf items not designed for biological use, and some may cause problems in the long run. Supplies of donor organs fall far short of demand; thousands of patients die every year waiting. Meanwhile, research on biology and materials advance apace. Capitalizing on science to address societal needs, a new discipline dubbed *tissue engineering* emerged in the late 1980s. Its vision is to produce living replacement parts for the human body, ready to be implanted.

Several approaches to tissue engineering have been proposed. One of the most promising is a prosthesis that coaxes and helps the body to heal itself by regenerating new living tissue. Such a prosthesis contains three major elements: living cells, growth factors and other biomolecules that signal and promote growth, and a scaffold to hold the cells and molecules before the patient's own new tissue takes over. Each element demands much R&D, as do the processes to fabricate and transport the prosthesis. Detailed knowledge about stems cells and their development is required to cultivate cells that can differentiate into the required cell type and avoid rejection by the body's immune system. Advanced biomaterials are required to make scaffolds that can harbor the cells, allow nutrients to reach them before blood vessels grow, contain drugs and biomolecules and release them according to controlled schedules, and gradually degrade to make way for regenerated tissues. Developmental biology and process engineering are required for designing bioreactors that can mimic the body's physiology and produce prostheses viable both before and after implantation. Several engineered tissue replacements have been approved by the U.S. Food and Drug Administration. They are mainly structural tissues—skin, bone, and cartilage—which are relatively simple because they do not require many nutrients. Off-the-shelf replacement hearts and livers are decades away, but tissue engineering is vibrant both as an industry and as a research area.[64]

The young fields of biomolecular, metabolic, and tissue engineering are full of vigor and potential. Drawing on biology, chemistry, physics, and engineering, they share a centripetal vision: to synthesize principles from various traditional disciplines into a coherent body of knowledge for predicting and manipulating biological systems on all levels, from molecules to organs. To achieve this synthesis, universities such as MIT and the University of California at Berkeley recently organized bioengineering divisions that mobilize the prowess of engineering analysis and quantitative modeling to tackle the complexity of biological systems. They will give birth to a new engineering science.

Recombinant DNA technology was developed by biologists, but it is aptly called genetic engineering because it is immediately useful in transforming nature and serving people. Bioengineering has a strong science base, but it does not forsake its utilitarian mission. Bailey remarked: "In addition to applying new knowledge from basic biology, chemical engineering will likely play a key role in the fundamental understanding of this exploding core of knowledge based on our experience with integration of information in a quantitative framework which describes the key features of a complex system. The field of biology itself should benefit significantly from chemical engineering's involvement in this sense."[65]

Curiosity and utility, the motivations of science and engineering, seem as far apart as the headwaters of the Mississippi and the Missouri. However, the two disciplines share the same real world and human mind, just as the two rivers share the same basin. Eventually they converge. For miles after the Mississippi and Missouri meet, the former's green waters and the latter's yellow waters flow distinctly from each other, but they are now in the same channel.

7

LEADERS WHO ARE ENGINEERS

7.1 Business Leaders in the Car Industry

Their sweat evaporating under the desert sun, a team of construction managers and civil engineers worked with Egyptologists to study the Great Pyramid of Giza. They were concerned less with physical methods than organizational ones, such as the logistics of construction and the sustenance of workers. The Great Pyramid's mass is roughly two-thirds that of the Hoover Dam. Its construction took an estimated average labor force of 13,200 workers more than ten years. How did the Egyptians orchestrate and provide for so many people? The team asked, How would *we* do it?[1]

Contemporaries of ancient master builders remarked on their managerial abilities in reckoning costs, securing supplies, coordinating efforts, communicating specifications, and resolving personnel conflicts. The care with which Callicrates chose contractors in erecting the Long Walls in ancient Athens impressed the Roman historian Plutarch, who described it in detail. Vitruvius explained that to construct eaves and sewers properly, the master builder had to be familiar with the laws of the city. Dealing with business and the government has always been a part of the engineer's job.[2]

In today's market economy and representative democracy, engineers have to deal with business professionals and consumers, government bureaucrats and interest groups. All these contacts require significant people skills. "An engineer has to know a lot about people, the ways they organize and work together, or against one another, the ways in which business makes a profit or fails to, especially about how new things become conceived, analyzed, developed, manufactured, put into use," wrote Bush. "The engineer stands in the middle of the spectrum between men and things, not only in his objectives, but also in the ways in which he uses science, and in the ways he operates in organization . . . In any engineering career there is likely to be a gradual shift from emphasis on knowledge of *things* to emphasis on knowledge of *men*."[3]

Engineering's human dimension, where engineers exercise their technical expertise in the roles of managers, policy makers, or simply responsible citizens, is no less significant than its physical dimension. Here purposive questions of what engineered systems are ultimately for, how well they serve their goals, and how to regulate their social impact come to the fore.

Another Stereotype Debunked

"Well-round and engineering school? Until recently, it would have seemed laughable to link the two phrases." Thus stated an article in *U.S. News and World Report* in 2000. Referring to MIT and other institutes "reputed to turn out antisocial brainiacs," it reported their recent efforts in "building a new breed of engineer—one with management savvy."[4] The news magazine rightly spotted the recent trend in engineering schools of stepping up education in business leadership, partly in response to student demand. It also revealed the prevalence of the nerd stereotype; only those blinkered by it would regard management savvy in engineers as a laughable idea.

Paul Litchfield, an MIT alumnus, wrote after becoming president of Goodyear Tire and Rubber: "Harvard men used to say the Tech would turn out engineers but that the nation would turn to Harvard for the head of business. Some of the men who were in school with me, however, got along all right, and I got a little satisfaction out of it as Alfred P. Sloan, class of '95, became head of General Motors, Gerard Swope, also '95, head of General Electric, Irenée du Pont, class of '97, was made head of that great company, and Roger Bobson, class of '98, became a world-famous economist."[5] This is not news; Litchfield was re-

Table 7.1   Percentage of top executives with an engineering background in large
U.S. corporations.

	1900	1925	1950	1964[a]
Percentage of top executives:				
Whose college degree was in engineering[b]	24.0	32.8	32.2	44.1
Whose first full-time job was as an engineer	7.7	12.2	16.8	26.6
Whose chief occupational experience was engineering	12.5	15.6	19.3	34.8

Source: M. Newcomer, *The Big Business Executive* (New York: Columbia University
Press, 1955), tables 27, 36, 38; Scientific American, *The Big Business Executive/1964,* tables
8, 14, 15.
a. 1964 numbers are for engineering or science.
b. Among those who held an undergraduate or graduate degree.

ferring to graduates of the 1890s. They were all educated as engineers.
Business schools at that time were merely struggling experiments.[6]

The significant involvement and success of engineers in business, as
revealed in the survey results of Table 7.1, would not surprise those
who attend more to work content than sociological stereotypes. Practical and hard-headed, engineers mean business. From railroads to automobiles to electricity to computers, new technologies have spawned
sunrise industries. Engineers with a mastery of those technologies have
an edge in exploiting their commercial potential. Nor are they incompetent in areas of management and finance. Economy and efficiency, although snubbed as vulgar by cultural elites, have always been valued
by common working people. Since the time of master builders, engineering projects have demanded skill in both weight and cost reckoning, both physical and organizational technologies. Coordination of
work becomes more important as technology becomes more complicated. Effective organization is indispensable in hard core engineering
design, such as airplane development projects that involve hundreds or
thousands of engineers. From this type of management, business administration is only one step away. It has offered engineers an attractive
route of advancement, especially in times before engineering science
took off and began to compete for talent.

Technical experts hold their own on the corporate ladder in competition with business schools' mass-produced MBAs. Of the seven corporations whose stories were featured in *Fortune*'s 2002 listing of the 500 largest American companies, two were headed by engineers and one by a scientist.[7] And this is just the tip of the iceberg, compared with the engineers who make up senior and middle management. A 1999 study by the National Science Foundation found that about 15 percent of those with a master-level degree in engineering become senior managers at some point in their career. The number rises to 35 percent for engineers who also hold a business degree.[8] Similar patterns for engineers in business obtain in Germany and Japan.[9]

Engineers make not only run-of-the-mill chief executive officers but reach the pinnacle of excellence. The two leaders in consolidating the entire German chemical industry during the interwar years were Carl Bosch, chairman of BASF, and Carl Duisberg, director of Bayer, two of the three top chemical firms. Bosch was an engineer, Duisberg a chemist. Both ascended from a technical research staff position to chief executive, and Bosch got a Nobel Prize on the way.[10] American engineers have been no less successful. Of the four finalists for *Fortune*'s "businessman of the century"—Henry Ford of Ford Motors, Alfred Sloan of General Motors, Thomas Watson, Jr., of IBM, and Bill Gates of Microsoft—three had engineering backgrounds. They represented a radical shift from businessmen of the nineteenth century. In *Fortune*'s words: "Organization Man rose to challenge the robber baron." *Fortune* also chose as "the manager of the century" Jack Welch, who, with a chemical engineering doctorate, started as a researcher in the materials department of General Electric, climbed to chief executive, and in the last two decades of the twentieth century strongly influenced management thinking by reengineering that lethargic conglomerate into a powerhouse.[11]

Fortune's two honors highlight two types of business talent: the founder of businesses and the professional manager, especially at the top level of large organizations. Ford, who was named businessman of the century, was an entrepreneurial founder, Welch a professional manager. Sloan, who first engineered the basic structures of giant multidivisional corporations, bridged the two.

Engineer-entrepreneurs are familiar today, since they have grabbed business headlines in the information revolution. They are only the most recent generation in a long heritage that will continue. Back in the first industrial revolution, Watt commercialized his steam engine and Stephenson his locomotive, and Maudslay and Nasmyth started the British machine-tools industry. In the second industrial revolution, Ford designed and produced the Model T, and he was only one of many Detroit engineers whose names appear on popular cars. More recently, Intel's first two presidents in the twentieth century, Noyce and Moore, both experimented with silicon themselves, and many other leaders in the semiconductor industry "had silicon under their fingernails," as electrical engineer Ian Ross put it.[12] Business is always risky, and success is never guaranteed, whatever one's background. Nevertheless, an entrepreneur hides inside many engineers, who know that what they make has commercial potential and who, with their can-do spirit, are less deterred by business risks.

Strategic Management of Large Corporations

At the end of the twentieth century, almost half of Americans owned stocks, either directly or through retirement plans. The democratization of ownership in enterprise capital requires skilled professional managers to run the corporations for diverse stockholders, most of whom are not involved in business. Guess who pioneered modern business administration? Combining their practical tradition with scientific reasoning, engineers were the first to be able to think big; think practically and objectively; think clearly, analytically, systematically. Inventing new approaches distinct from those of traditional merchants, they introduced strategic, operational, and tactical management in the large corporations that began to emerge in the mid-nineteenth century.[13]

Officer and military scientist Carl von Clausewitz defined two kinds of activity: "that of the *formation* and *conduct* of these single combats in themselves, and the *combination* of them with one another, with a view to the ultimate object of the War. The first is called *tactics,* the other *strategy.*"[14] Replace "combat" and "war" with "productive activity" and "business," and the same statement describes tactical and strategic business administration. Strategic management establishes corporate architecture, designs business models, anticipates market demands, plans product mixes, allocates resources to and coordinates efforts of various operations, evaluates performances, and adjusts plans, all with long-term objectives, such as continuously maximizing return on investment. Tactical organization arranges finance, chooses and integrates manufacturing processes, coordinates workers and machines, arranges supply logistics, and decides on specific products, with criteria such as improving productivity. Operational organization aligns tactics to the strategy. Each depends crucially on relevant intelligence and is responsible for designing and maintaining efficient channels of information flow.

In the old days, when business firms were small and their operations simple, family management sufficed. As business scale and complexity escalated in the late nineteenth century, corporate organization became increasingly crucial. Strategic and tactical management were respectively pioneered by civil and mechanical engineers, who took on the new jobs to meet the developing challenges in their productive processes. Business historian Alfred Chandler wrote: "The most creative of those men who laid the groundwork for modern American corporate administration were three professionally trained engineers—Benjamin H. Latrobe, Daniel C. McCallum, and J. Edgar Thomson. Of the many contenders for the title of the founder of modern business administration, these three have the strongest claim."[15] All three were civil engineers who built railroads and later took to managing the operations of those railways.

Railroads, the first enterprise of sufficient scale operating at sufficiently high speed, brought a new kind of business that befuddled old-fashioned management. By the 1850s, the large rail lines were operating hundreds of locomotives and thousands of freight cars that carried loads millions of miles per year. Prompt and efficient operation of such complex systems, with numerous operators and agents and a haphazard flow of money, demanded a novel administrative approach. Civil engineers naturally filled the job, not only because they were familiar with the physical technology but also because they were already experts in planning, organization, and logistics.

Working independently on three railroad lines, Latrobe, McCallum, and Thomson applied their rational and analytic minds, honed in engineering practice, to solve the new managerial problems. They created the first elaborate internal organizational structures appropriate for private enterprise. Their organizations included separate central office and regional divisions, distinct functional departments, clearly defined loci of authority and responsibility, and efficient channels of information flow. For the first time, information gathered within a corporation was used extensively to aid coordination and planning. With these innovations and "general principles of administration," they developed groundbreaking models for the structure of modern large corporations. Through the intermediate influence of financiers, their management methods became models for other businesses.

With the advent of mass production in the early twentieth century, giant manufacturing corporations appeared. Manufacturing was more dynamic than railroad operation and required more sophisticated administration. Prominent among those who completed the managerial revolution started by railroad engineers was a new generation of engineers who established the basic structures of American corporate capitalism: Sloan; Swope, who diversified the product lines of GE; Alfred, Coleman, Irenée, and Pierre du Pont, who reorganized a sleepy family company and turned it into a galloping giant. They were college-educated engineers in various fields, sometimes unrelated to the business of the corporations they came to head. Sloan was educated as an electrical engineer, although he spent his career in the automobile industry. What made them innovative executives was not specific pieces of engineering knowledge but their general technical aptitude and scientific approach to handling complex situations. Sloan wrote: "General Motors is an engineering organization . . . Many of our leading executives, myself among them, have an engineering background."[16] Applying the engineer's practicality, rationality, thoroughness, and scientific approach to financial control and corporate governance, those engineers designed a corporate structure at GM that became the model for large multidivisional corporations. "Reengineering corporations," a term popular in the last decade of the twentieth century, is apt because corporate structures were engineered in the century's early decades.

Industrial Engineering and Tactical Management

A corporation contains many functional arms, such as finance, manufacturing, marketing, and R&D. Coordinating the operations of these arms falls on strategic managers. Operating each arm calls for managers who are more familiar with functional substance. Engineers have become leaders in developing systematic management of manufacturing or more generally physical production.[17]

To rationalize and systematize production processes, mechanical engineers are generally in a better position than civil engineers. Efficiency of construction processes is difficult to improve on because the processes must adapt to local conditions and rely on local supplies that differ greatly from construction site to construction site. Most logistic schemes for building the Hoover Dam, for example, were inapplicable to the Aswan High Dam because Colorado and Egypt had different local conditions that civil engineers could not change. In contrast, manufacturing takes place in fixed factories with persisting factors, which can be controlled and optimized for efficient operation, gradually if not immediately.

Starting in the 1880s, *American Machinist, Engineering News, Engineering Magazine,* and *ASME Transactions* published articles on how to organize factory floors, select manufacturing processes, measure costs of human and machine operations, coordinate and monitor their activities, design wage-payment and incentive plans, and synchronize smooth flows of supplies. Management was the theme of the 1886 annual meeting of the American Society of Mechanical Engineers (ASME). Henry Towne decried the lack of planning in factories and urged engineers to develop methods for taming the chaos. His talk was followed by others addressing problems in cost and capital accounting. In 1920 the ASME established a management division, which quickly became the largest division of the society.

University courses in industrial management first appeared in 1904 at Cornell's Sibley College of Engineering. A comprehensive study conducted by the Society for the Promotion of Engineering Education in the 1920s recommended that economics courses be required for engineering students.[18] These requirements stemmed mainly from the substantive content of effective production and not social contexts idiosyncratic to America. German technical institutes also introduced economic and managerial subjects into their engineering curriculum, culminating in the creation in 1921 of a special major, economics-engineering. This was the beginning of *industrial engineering,* an engineering management discipline aiming to improve productivity by efficient coordination of human, material, physical, and financial resources in productive operations.[19]

From "The American System" to Mass Production

For examples of engineers in business, let us take a whirlwind tour through the history of the machine-tools and automobile industries. In most modern manufacturing processes, the parts of a complex product are made separately, then put together in final assembly. Even today, when interchangeable parts are taken for granted, assembly is still a major operation that on average takes up about 20 percent of unit cost and 50 percent of total production time.[20] Lowering costs by improving assembly efficiency is paramount in manufacturing. Historically, the most important step forward was the introduction of mass-produced interchangeable parts, which depends on integrating physical and organizational technologies.[21]

By itself, fabrication of identical parts is precision craftsmanship. It makes an economic impact only under a business strategy that demands production of complex products in sufficiently large quantities, at high speed and low cost. Before the rise of modern industry, the military was the party that would benefit most from large-scale production and consumption of standardized equipment. Aiming to rationalize French artillery and other military practices based on scientific reasoning, General Jean-Baptiste de Gribeauval developed a broad strategy called the *uniformity principle*. Under it, he supported Honoré Blanc in tooling a government arsenal with die forging, jig filling, and other techniques. Blanc's muskets with interchangeable parts impressed many people. Among them was U.S. ambassador to France Thomas Jefferson, who assembled several muskets with his own hands and enthusiastically wrote home about it in 1785.

Gribeauval's program withered in the political storms of France but bore fruit in America, although not through the hype of Eli Whitney.[22] Blanc refused Jefferson's invitation to come to America, but several French engineering officers joined U.S. service. Among them was Louis de Tousard, who elaborated the uniformity principle in a 1809 military textbook written at the behest of George Washington. In 1815 Congress authorized America's two national armories to make regulations "for the uniformity of manufactures of all arms ordnance." This mandate was doggedly pursued over the following decades. Assuming leadership was the Springfield Armory, under superintendent Roswell Lee. On the strategic level, Springfield rationalized its product mix by slashing the number of calibers and developing only a few models for production. On the operational level, it standardized bookkeeping and orchestrated similarity in production processes by designing jigs, fixtures, and gauges. In the process, Springfield made itself into what Chandler called "a prototype of the modern firm where modern factory management had its genesis."[23]

In 1854 the British came shopping for rifles with interchangeable parts, which they called "the American system." Britain had just vaunted its technological might at the Crystal Palace Exhibition, but its weak spot was revealed in its need to import technology for national security from a former colony on which it had imposed a strict technological embargo. Skilled British workers could make rifles with higher *quality* than the Americans but could not achieve *uniformity*. Quality is an attribute of individual rifles and calls for skilled individual workers. Uniformity is an attribute of large batches of rifles, and achieving it requires organizing the efforts of many workers and developing the physical technology to combine their work. Integration of organization and physical technology in production was the American strong hand that later made modern American corporations of mass production so formidable.

Besides developing and systematizing production processes, Springfield Armory became a technology clearinghouse for the private arms contractors doing business with it. It helped many start-ups and in return benefited from their inventions. An example was Robbins and Lawrence of Vermont, an arms and machine-tools firm formed in 1845 with little capital but much engineering brilliance. With products admired by the British, it shot to international fame in six years, opened the export market for American industry, expanded too quickly, overstretched, and collapsed in another six years—a story not unfamiliar in the history of technological ventures. Nevertheless, it left its mark. Many other companies did last, among them Samuel Colt's armory, built in 1853.

The armory community spawned many able engineers who moved

on to typewriters, sewing machines, bicycles, and other industries, bringing the production technology with them. An outstanding example was Henry Leland, who started his career designing machine tools at the Springfield and Colt armories. After a sojourn in sewing machines, he went to Detroit and founded Cadillac in 1904 and the Lincoln Motor Car Company in 1917. His Cadillac was the first car to be built completely from interchangeable parts. Once Sloan, a green college graduate working as a sales engineer for a roller bearing company that supplied the car industry, went to see Leland about certain complaints. Leland whipped out a micrometer, showed Sloan that his bearings did not meet specifications, and told him: "Young man, Cadillacs are made to run, not just to sell." Reminiscing in his late eighties, Sloan wrote: "[Mr. Leland] was a generation older than I and I looked upon him as an elder not only in age but in engineering wisdom."[24] He traced through Leland's career the "line of descent" for the automobile industry. The line was long indeed.

Mass Production in the Automobile Industry

Two events shook the American automobile industry in 1908. Ford rolled out the Model T, and William Durant put together General Motors by merging Leland's Cadillac with the companies founded by engineers David Buick, Ransom Olds, and Louis Chevrolet. Thus began a long duel between the Big Two that would end with each adopting the other's strong suit. The two major antagonists, Ford and Sloan, set a model for giant manufacturers with their pioneering contributions to tactical and strategic management.[25]

Ford was an engineer more of cars than of manufacturing processes. To design the plants and operations required for his business strategy of expanding demand by lowering price and hence manufacturing cost, he gathered a team that pioneered industrial engineering. They systematically developed the technologies of jigs and standardized measures accumulated since Springfield. Furthermore, they integrated machining steps into work sequences, arranged machines and tools, laid out supply routes for parts, choreographed worker activities, all with eyes on enhancing productivity. Their crowning triumph came in 1913, when moving assembly lines started to bring parts to stationary workers, ensuring a smooth flow from parts to finished products. Output nosed up and price down. In the peak year of 1923, Model T production reached 2 million while its price was one-eighth of that in 1909, adjusted for inflation.

General Motors was choking in the exhaust from Model T's tailpipe. An adroit financier but a lousy manager, Durant was forced out when GM faced bankruptcy in 1920. Taking over the mess was a team of four engineers headed by Pierre du Pont, who vacated the presidency for Sloan two years later. They introduced systematic management, distinct from Ford's entrepreneurship.

Ford organized his manufacturing operation to the last detail, but allowed no mechanism to steer this gigantic operation except his own genius. Human genius is susceptible to erosion by age and corruption by power. Sloan wrote of Ford and Durant: "They were of a generation of what I might call personal types of industrialists; that is, they injected their personalities, their 'genius,' so to speak, as a subjective factor into their operations without the discipline of management by method and objective facts."[26] Sloan provided objective discipline by instituting what Ford loathed, a corporate architecture with top-level management capable of stable yet flexible business strategies.

Sloan's move was two-pronged: strategy and organization. For strategy, he decided on the product mix of "a car for every purse and purpose" with "constant upgrading of product." To ensure the viability of this dynamic strategy, he designed a corporate structure called "decentralization with coordinated control." Engineers would recognize in its basic principles the *abstraction, modularization,* and *interface* familiar in their systems approach. In the corporate context, modularization takes the form of divisional organization, and abstraction takes the form of management by numbers. With modularization, Buick and other divisions run their operations autonomously. Their autonomy is not absolute; they are all subject to the policy of a central headquarters, which demands reports and forecasts from them, evaluates their performance, and coordinates their efforts through appointments and resource allocation. The required reports, which serve as the crucial interface among various parts of the corporation, contain only essential numbers and factors. By abstracting from the details of operation, they save top management from information overload and the divisions from micromanagement, leaving operational decisions to those who know the local situations best.

It was a gigantic task to design and implement an organizational structure with adequate provisions for information flow, financial control, standardization that accommodates divisional autonomy, marketing and distribution, and research and development. It took Sloan many years, and he finally prevailed. In manufacturing operations, GM adopted most of Ford's mass production tactics to suit its strategy of planned obsolescence. For its part, Ford Motors imported wholesale GM's organizational architecture when Henry Ford vacated his office in 1945.

Ford and Sloan, engineers of two generations, were rivals but also complements of each other. To Ford's innovations in product, manufacturing, and industrial engineering, Sloan added corporate engineering for a manufacturer's flexibility and stability in expanding, diversifying, and weathering storms. Together they created the model for a viable giant corporation for mass production, giving life to the epithet "American system of manufacturing."

From Mass Production to Lean Production

The year 1950 was the heyday of mass manufacturing. America produced more than 6.7 million cars, which accounted for 76 percent of world production and captured 98 percent of the domestic market. Japan produced a total of 2,396 passenger cars.[27]

That year Eiji Toyoda, Toyota Motors' head of engineering, visited Ford Motors. His host asked what he wished to learn. "Quality control, production methods, the works," he answered. "You want to know too much," the Ford official said; "besides, nobody in Ford knows everything that goes on around here." "Maybe you don't have anyone like that at Ford," Toyoda thought, "but we do at Toyota."[28]

No, Toyoda was not referring to the comprehensive systems thinking that was budding in the American aerospace industry. He came from a very different culture. However, the technical problems he faced in manufacturing were not so very different. Many industrial practices that Toyota later developed would become integral parts of systems and concurrent engineering as discussed in section 5.2. But that would be decades later.

Toyoda inspected Ford's factory operations, asked questions, learned a great deal, and was heartened: "I didn't think that there was such a large gap in technology." What he saw in place was two decades old. Ford's new managers imitated Sloan's corporate organization but did not heed his advice that "knowledge of the business is essential to a successful administration."[29] They were not car builders but bean counters, more interested in maximizing short-term profit than investing in technology. They did not get an honorary mention in *Fortune's* "businessman of the century" article. Toyoda did. He was what Detroit calls a "car guy." After getting his mechanical engineering degree, he performed hands-on jobs in all aspects of car design and manufacturing, starting from reverse engineering, where he tore apart American cars to study their strengths and weaknesses. With his technical experience, he could see the glaring waste in Ford's factories that escaped Ford managers frantic about cutting costs.

Back in Japan, Toyoda convinced skeptical Toyota directors that they could compete with American automakers. He initiated a strategic plan to rationalize production processes and chose as his lieutenant Taiichi Ohno, an industrial engineer with a broad vision: "Efficiency must be improved at each step and, at the same time, for the plant as a whole."[30]

"Those who know both themselves and their opponent win a hundred times out of a hundred," Sun Tsu wrote in his *Art of War.* The Toyota team studied American production methods thoroughly and identified at least two major sources of waste. The first lay in the practice of having each work station crank out parts as fast as possible. This produced stockpiles of idle inventory that waited for assembly, crowded factory space, tied up cash flow, drained accounting resources, and risked the possibility of obsolescence in design changes. To target inventory waste, Toyota developed methods for *just-in-time production:* producing just the right number of parts at the right time, when they are needed for assembly. This is a difficult goal. A producer of car parts, such as the transmission, in turn needs smaller parts and must manage its inventory to deliver on demand. If the assembler simply dumps the cost of stockpiling on suppliers of parts, it leaves the suppliers no alternative but to jack up parts prices overtly or covertly, thus foiling the whole purpose. To realize the efficiency of just-in-time production, proper coordination must be established between the assembler and its suppliers, between the suppliers and their suppliers, and so on all the way down the supply chain. Today the design and management of supply chains are hot topics in engineering and management schools.

The second area of waste lay in placing quality control only at the inspection of finished products. Most problems were caused by defective parts or parts improperly installed. Once defects were embedded in a product, they were hard to detect and rectify, making quality control difficult and reworking expensive. Ford employed an army of quality control managers. Their ineffectiveness was pointed out by a perceptive assembly worker: "You have inspectors who are supposed to check every kind of defect. All of us know these things don't get corrected . . . I can just look at a car and see all kinds of things wrong with it. You can't do that because you didn't know how it was made. I can look at a car underneath the paint."[31] The valuable experience of production workers may have escaped the bean counters who had never set foot in the factory but not Ohno, who had risen from the ranks. To target reworking waste, Toyota introduced *total quality control*: nip trouble in the bud by using the peculiarity of assembly and the experience of assembly-line workers. Fix any defect while the product is on the assembly line, and root out the causes of the defect so that it does not happen again. Defects are more visible to assembly-line workers, who are familiar with car parts and see them exposed in unfinished cars. For total quality control Toyota asked the workers to fix any defect they noticed—to stop the whole assembly line and get help if necessary—so that no spotted defect slipped through. Moreover, engineers worked on the factory floor to trace any trouble spot back five causal steps and make corrections to prevent a future occurrence. Assembly workers and engineers directly added value to the products. When they took over the responsibility of expensive quality control specialists, cost dropped while quality rose. Toyota became the world's third largest automobile manufacturer in terms of units by 1971. A decade later

Japanese officials would bow to American visitors and ask what they wished to learn.

Japan succeeded in many industries besides automobiles and posed a great threat to American industry in the 1980s. Many factors underlay the phenomenon. Refusing to accept elitist theories that it was all because of Japan's inscrutable culture, American engineers and managers analyzed practices in the two countries, compared relative advantages, abstracted from cultural idiosyncrasies, and delineated important methods and principles that they called *lean production,* in contrast to *mass production.* Once they figured out the integration of organizational and physical technologies that made lean production work, they quickly adapted and implemented them in their own businesses, turning the tide on the Japanese. Leanness—manufacturing high-quality products with minimum labor and capital—has spread from automobiles to aerospace and other manufacturing and has become integral to industrial engineering and management.[32]

Lean and Mean?

Mass production, with its minute division of labor, created a new breed of workers whose plight was poignantly satirized in Charlie Chaplin's *Modern Times.* In comparison, lean production's promoters claim that it enables workers to think more creatively and master more skills, while critics claim that it is management by stress and thus is far more exploitative. Careful empirical studies have discredited both extremes. Increasing automation in both mass production and lean production relegates most humans to the "residual work" of tending or feeding machines. Their tasks are narrow, fragmented, choreographed, standardized, and timed, with an average cycle time of one minute. Because lean production makes the most of workers through "constant improvement" and other practices, its pace is indeed more relentless, as critics charge. At NUMMI, a joint venture of GM and Toyota that practices lean production, workers put in 57 seconds of labor every minute, compared with about 45 seconds per minute when the plant was run by GM alone.[33]

Nevertheless, NUMMI jobs are highly sought after, and most NUMMI workers say they are satisfied. Lean producers use sociology and psychology to compensate for their workers' quicker pace. They organize workers into small teams and train them in team dynamics, so that they are under peer pressure to perform and in return receive social support from their teammates. Lean manufacturers also institute job rotation or multitasking. Where five mass production workers each engage in a one-minute task, four lean workers accomplish the same job by each performing all five tasks. Employers gain in saved labor cost and a more flexible workforce, and a part of that gain is passed on to consumers in the form of lower prices. Employees get reduced monotony in return for a heavier workload, a trade-off many—although not all—say they welcome. Lean producers try to align worker incentives with corporate goals and to drum up workers' sense of pride and self-importance. Workers are urged to make suggestions and thus gain a sense of autonomy, although almost all important decisions are made at the top. They are reminded of their achievements in producing the high-quality products that make their employer a world-class corporation. The value of such psychological satisfaction cannot be underestimated. On the whole, leaner is not meaner, but neither is it kinder and gentler.[34]

Few want to squeeze and torture fellow human beings. People just want to maximize their own benefits. The question is, How much do the powerful care about the potentially devastating side effects of their pursuits? Ohno wrote that he was "advocating profit-making industrial engineering . . . Unless IE results in cost reduction and profit increase, I think it is meaningless."[35] Undoubtedly profit from cost reduction is a major aim of industrial engineering, but is it so important as to crowd out other factors, such as employee relations? Should it be? These questions have been debated by engineers ever since the first days of professional societies. Engineers must organize both human and physical resources. As engineering projects become more complex and involve more factors with often conflicting claims, engineers' social responsibility increases. Responsibility comes with leadership.

7.2 Public Policies and Nuclear Power

As technology now permeates the social fabric, it has entered many public policies and legal proceedings. Technical expertise is required for everything from negotiating international nuclear test ban treaties to adjudicating whether a worker was hurt by factory radiation. To fill this need, more than twenty American engineering schools have departments or centers that combine engineering with law or public policy. Integrating technical and social sciences, these programs extend the practical attitude and analytic ability of the engineer to addressing problems that have broad social impacts. The Engineering and Public Policy Department at Carnegie Mellon University aims to "advance the state-of-knowledge and -art in how engineering policy problems are formulated, solved and interpreted for policy insight and development." The Technology and Policy Program at MIT strives to produce not only "engineering leaders" but "leaders who are engineers," skilled intellectually to deal with the many crucial technological dimensions of our society. This is perhaps the hardest task for engineers, because here they are fully exposed to the responsibilities of personal, professional, and public ethics.

Engineering and policy programs usually have a global scope. Policies in areas such as telecommunications necessarily have international dimensions. Technology policies play an important role in economic development, and even national policies must be viewed in the context of globalization. Research at Carnegie Mellon, for instance, emphasizes policies in China and India. Policies and political processes differ widely among nations. For brevity here I will focus on the United States.

A policy is a plan or strategy designed to guide, influence, or determine specific decisions or actions. Public policies that somehow involve science and technology (S&T) fall into two overlapping classes. The first, known as *S&T policies,* address strategies for or about S&T. The second, *S&T-dependent policies,* address such matters as environmental protection, arms limitations, or industrial structures that are not about S&T but require significant input from them.[36]

The rationale for S&T policies is enunciated in the first two principles of the President's Committee of Advisors on Science and Technology: "1. Science and technology have been major determinants of the American economy and quality of life and will be of even greater importance in the years ahead. 2. Public support of science and technology should be considered as an investment for the future."[37] To make the investment judiciously, policy makers try to identify national goals; coordinate education, research, and development; select nascent areas, such as information technologies and health research, that require nurturing before market forces can take over; evaluate the effectiveness of previous programs; propose budgets adequate for achieving the multidimensional goals. More specifically, science policies emphasize basic research with long time horizons. Technology policies aim to enhance the effectiveness of technological innovations in serving the nation's need for high productivity, economic competitiveness, environmental quality, and national security.

Standards Setting as S&T Policy

An area of S&T policy that is more important than glamorous is standards setting. The advantages of standardized weights and measures in facilitating commerce and political unity have long been appreciated. With science discovering ever more quantities and technology inventing ever more devices, standardization becomes more complicated and vital. Standards come in all levels of generality, from scientific units such as the ohm to the compatibility of particular systems such as the threads of nuts and bolts. Consumers who take for granted plug-and-play gadgets and international telephone calls should be reminded of the 1904 Baltimore fire, which razed more than 1,500 buildings while reinforcement firefighters from neighboring cities and states watched and wrung their hands because their hoses did not fit Baltimore's hydrants.[38]

Development of successful standards requires a tremendous amount of skilled negotiation to forge consensus and procedural care to ensure fairness and openness. Many governments have departments for setting standards. The European Union established three official organizations for European standards. The United States is peculiar for its decentralization. The federal government decrees regulations but leaves most standards development to the private sector, participating only as one of many parties. Today standards are set in the United States by some 600 organizations that include government agencies, trade associations, professional societies, testing institutes, and corporate consortia. Among them, engineers have a long tradition of being in the lead.

Engineers aim to make things work, and standardization is required for many things to work effectively. As soon as American engineering societies appeared, they formed committees to develop standards. At first standards were confined to individual industries or local areas. In 1918 societies of electrical, mechanical, mining, and materials engineering came together and invited three government agencies to join them and establish what is now the American National Standards Institute (ANSI). It quickly admitted other parties and grew into the chief organization for coordinating America's standards development activities and approving standards on the national level. The ANSI is also the sole U.S. representative in the two most important international standards organizations, a crucial role in the age of globalization.

Standards can systematize past experiences or project future developments. Retrospective standards distill the knowledge embodied in a matured technology and facilitate its promulgation. ASME's Boiler Code, first published in 1915 and continually revised since, systematically spells out the requirements for the safety of pressure vessels. Anticipatory standards, initiated before a technology matures, evolve together with it, encouraging innovation and widespread acceptance. The TCP/IP protocol suite, developed and maintained by the Internet Engineering Task Force, was instrumental to the rapid expansion of the Internet.

Retrospective or anticipatory, standards as specifications accepted by diverse parties have significant policy ramifications. They are crucial to the proper functioning of society's infrastructures and strategic in channeling technological progress. Their proper development and maintenance impose heavy social responsibilities on engineers. The ASME was convicted for antitrust behavior in 1982 because it did not prevent a conflict of interest when members interpreted a clause of its Boiler Code in a way that deterred the introduction of an innovative valve. Upholding the verdict, the U.S. Supreme Court wrote: "ASME wields great power in the nation's economy. Its codes and standards influence the policies of numerous states and cities, and as has been said about 'so-called voluntary standards' generally, its interpretations of guidelines 'may result in economic prosperity or economic failure, for a number of businesses of all sizes throughout the country,' as well as entire segments of an industry."[39] Chastised, the ASME thoroughly revised its procedures for handling codes and standards.

Economic and Social Policies Requiring S&T Input

Most policies focus not on science and technology but on economic, geopolitical, and social issues. Some of these issues are induced by new S&T, as information technology and genetic engineering spawned new issues of intellectual property rights. The resolution of other issues requires instrumental roles of S&T, as policies on global warming depend crucially on scientific knowledge and technological feasibility. The bulk of S&T-dependent issues fall under regulatory policies, which divide roughly into two groups: economic and social.[40]

At the beginning of the twentieth century, physical technologies in telecommunication and transportation and organizational technologies in large-scale production and distribution ushered in the new mass-consumption economy. To cope with its social ramifications, which were accentuated by the market failure of the Great Depression, a flurry of legislation around 1933 put telecommunication, airlines, banking, utilities, and other industries under regulation. Although they are called economic regulations, many had technological constraints and ramifications. Despite strong antitrust sentiment, regulation of telephone service as a monopoly held its bite until new technologies enabled radical deregulation with the Telecommunications Act of 1996, which unleashed a restructuring of the telecommunications industry. Designing policies for integrating a host of technologies into a global information infrastructure in a deregulated economy is among the top agendas in the technology and policy programs of many engineering schools.

Rising consciousness of civil rights and environmental issues in the 1960s brought on a tide of social regulations designed to protect the health and safety of citizens and workers; to prevent harm from pollution and toxic substances in land, air, water, and workplaces; to promote civil rights and ethical objectives. Congress created the Environmental Protection Agency (EPA) in 1970 and the Occupational Safety and Health Administration (OSHA) a year later, delegating ambitious programs to both. Since then the number of social regulatory statutes has exploded.

To make laws that are effective and fair to citizens, a democracy tries to ensure that the law-making processes are open, so that all parties are given adequate opportunities to present their facts, arguments, and opinions. First comes *agenda setting,* when numerous problems jostle for government attention. An issue that makes the agenda undergoes the process of *legislation,* during which the new regulation is developed and formulated clearly, debated by interested parties, and, if it wins enough support, enacted. Legislation usually prescribes only broad goals and the strategies for achieving them, leaving details to the discretionary power of an administrative agency. To *implement* the policy, the agency makes substantive rules, enforces them, and evaluates their effectiveness, following certain decision-making procedures that emphasize due process, transparency, and stakeholder participation.

Technical ideas can be influential in both the legislation and the implementation stages. In many cases, the devil is in the details of implementation. To get information, besides partisan reports, government departments and agencies maintain their own technical staffs or advisory committees, hire consultants, and seek advice from each other and academia. After the nuclear reactor accident at Three Mile Island, for example, Congress requested its advisory Office of Technology Assessment (OTA) "to assess the future of nuclear power in this country, and how the technology and institutions might be changed to reduce the problems now besetting the nuclear option." The OTA assembled an advisory panel and organized workshops that convened representatives of all parties, including engineers and scientists. Its comprehensive 1984 report helped to shape the Energy Policy Act of 1992, which introduced reforms to alleviate some problems of the nuclear industry.

Many Facets of Policy

It is through agency staff positions and advisory channels that engineers and scientists have the most leverage in policy making. "Policy makers are always hungry for new ideas that they can turn into popular legislation," says professor of engineering and policy Jon Peha, who encourages engineers to participate in policy making. And he adds: "To be useful, suggestions must take into account political as well as technical realities."[41] In public policy on information and communication, energy and the environment, technical factors are interwoven in a com-

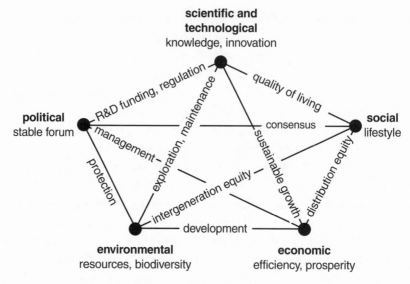

Figure 7.1 Five interrelated dimensions affecting policy for sustainable development.

plex of economic concerns, social values, and political interests (see Fig. 7.1). Any specialist would be baffled by the spectrum of positions on a controversial political issue.[42]

Effective policies take into account as many factors as possible. Science and technology—their potentials and limitations—are to be taken seriously, not just seen as magic wands or ideological scapegoats. The economic and technological perspectives are alike in their respect for efficiency. Economic analysis, which reckons with aggregate costs and benefits of options for prosperity, pays scant attention to the distribution of costs and benefits. Distributional equity belongs to social issues, which also include forces of culture and lifestyle. Environmental concerns compel economic analysis to internalize costs of pollution and environmental degradation. By advocating conservation of natural resources and long-term husbandry of ecosystems, they promote the welfare of future generations and hence a sense of intergenerational equity. Claims of various perspectives pull in different and sometimes opposite directions. Political institutions provide stable arenas wherein the diverse parties can come to some consensus on actions, but their procedures impose additional constraints on choices. A good example of how the S&T, economic, social, political, and environmental forces affect policies for engineering systems is seen in the nuclear power industry.

The Rise and Fall of Nuclear Energy

If defeat is a better teacher than victory, then the nuclear electric power industry has plenty of lessons to offer. In twenty-five years, President Dwight Eisenhower's "atoms for peace" program turned into activist Ralph Nader's "technological Vietnam."[43] Three Mile Island and Chernobyl top the list of Frankenstein's monsters, and nuclear energy has become an ideological whipping boy symbolizing "invasive new technologies to trample popular wishes." Thoughtful analysis, however, has revealed failures in the intricate interplay among many factors. The lessons learned are timely in a world facing difficult decisions regarding global warming.[44]

Worldwide, 436 operating nuclear power plants in 31 countries were joined by 2 newcomers in 2001. Together they generate about 17 percent of the world's electricity. Another 32 plants are under construction, mostly in Asia. The status of nuclear power varies widely among nations at similar technological levels. Germany is under political pressure to phase it out, while neighboring France happily depends on it for 76 percent of its electricity. In the United States, the amount of electricity generated by nuclear energy has risen in pace with the nation's rising electricity consumption, contributing steadily more than 19 percent from 1990 onward. This variety among nations reveals the hands of policy.

Nuclear power began in 1942, when physicist Enrico Fermi demonstrated that the energy released by nuclear fission could be harnessed in controlled ways. When World War II ended, Congress established the Atomic Energy Commission (AEC), among whose missions was the development of power-generating nuclear reactors. The first aim was submarine propulsion. An intensive project under Admiral Hyman Rickover culminated in the nuclear submarine *Nautilus*. It demonstrated that nuclear reactors could be designed, constructed, and operated safely.

A few weeks before the launch of *Nautilus,* President Eisenhower announced the "atoms for peace" program, and seven months later he signed the Atomic Energy Act of 1954. It declassified much technical information and added to the AEC the dual responsibilities of promoting and regulating commercial generation of nuclear power. Adapting technologies for pressurized water reactors originally designed for submarines and aircraft carriers, the government cooperated with a private utility company to build a demonstration power station at Shippingport, which was connected to the electric power grid in 1957. With a capacity to produce 60 megawatts (MW) of electricity, it was comparable to a middling coal-fired power plant of that era, but it did cost more.

The initial success fanned a euphoria in the government. The only group opposing rapid commercial deployment was the AEC's own Reactor Development Division. Scientists and engineers in reactor research were as enthusiastic about nuclear energy as anyone, and in the long term they were its staunchest supporters. Being more knowledgeable about its technological complexity, however, they argued for more time to develop reactor designs with better prospects for economic viability. Their caution was dismissed as "perfectionism" by politicians, who also suspected, correctly, that the Soviets were already building them. The race for megawatts was on. The consequence of this technological rush would be a fall as swift as the rise of the nuclear power industry.

The utility companies that invested in and operated power plants were not used to high technology. Nuclear reactors to them were just another way to boil water. Enticed by the government, reactor vendors, and a forecast of rising electricity demand, they began in 1965 to place a flood of orders for nuclear power plants. The bandwagon market was peculiar in its variety of designs and the rapidity of scale up. These technical mistakes, coupled with the weakness of policy tools and the slackening of electricity demand, cut the boom short. All orders placed after 1973 were cancelled.

Industries usually strive for standardization to enhance efficiency. The Canadian and French nuclear power industries operate efficiently with only a few standard reactor models. In contrast, almost every nuclear power plant in the United States is one of a kind. Each utility company built a reactor or two to suit its particular needs, and the AEC was unable or unwilling to coordinate any standardization. The variety of plant designs required custom-built safety systems; lengthened licensing reviews; multiplied uncertainties in regulation and performance; increased costs of design, finance, manufacturing, maintenance, and personnel training; and impeded the transfer of knowledge and experience from one reactor to another.

Many nuclear power plants ordered in the 1960s were behemoths with capacities exceeding 1,000 MW, more than four times the size of the largest working plants. The leapfrogging increase in plant size, meant to achieve an economy of scale, strained the learning process in developing complex new technology. The industry and the AEC adopted a "design-as-you-go" approach, the opposite of systems engineering. Plant construction would begin when a design was only 15 percent complete, and it would be modified whenever new experience brought new regulations. The constant stream of changes to plants in operation or under construction added enormous costs, especially interest costs caused by delay. Instead of inexpensive power, some—although not all—nuclear plants gave their customers rate shocks.

Some changes did more harm than good by interfering with other components in the complex system. By the late 1960s, technical staffs in the AEC and at the Oak Ridge National Laboratory increasingly complained that the AEC put safety second to the industrial boom. Their discontent, leaked to a public already anxious about fallout from atmospheric nuclear weapon tests, added to concerns about radioactive waste and the possible leakage of low-level radiation. Popular support for nuclear power, which had been overwhelming, softened in the 1970s and crumbled after the accident at Three Mile Island.

Congress in 1974 changed the AEC into the Nuclear Regulatory Commission (NRC), shorn of any mission to promote nuclear power. In the 1980s, many people proclaimed that the American nuclear power industry was dead or would die when operating plants shut down at the end of their forty-year licenses, if not sooner.

Nuclear Energy in Energy and Environmental Policies
Amid the gloom, the industry struggled to pull itself together. The Energy Policy Act of 1992 helped. Directly, it instructed the NRC to streamline licensing processes and lighten regulatory burdens. Indirectly, its deregulation of the electricity industry engendered an indus-

trial restructuring favorable to nuclear power. Competitive rationalization weeds out poor performers and consolidates ownership for efficient operation. Instead of a large number of utility companies each running a nuclear reactor or two, now large operators run fleets of reactors. They have combined personnel and expertise, standardized maintenance procedures, reduced down time and refueling time, and streamlined management. Safety has improved. Operating costs dropped and productivity rose. During the 1990s, the number of operating reactors declined from 112 to 104, but the electricity they produced increased by 30 percent. Now nuclear is the least expensive energy in terms of fuel, operation, and maintenance.[45]

As the century turns, business is again brisk at the NRC. Operating plants are filing to renew their licenses for up to twenty years or to expand their capacities up to 15 percent.[46] Talk about new plants, once deemed impossible, is being heard. Hope for a revival, however, is still immature. Nuclear power must overcome many hurdles, the highest of which is economic. Capital costs per unit of electricity for new nuclear plants are much higher than for natural-gas-fired plants. Unless the price of natural gas shoots up, nuclear energy is not financially attractive.

At nuclear energy's nadir, the OTA's 1984 report found that although it was not necessarily vital, "there may be good national-policy reasons to see the nuclear option preserved."[47] At issue is not merely the popularity or financial predicament of a single industry. Nuclear power must be assessed in the context of the nation's policy on energy and the environment as a whole, which is in turn linked to economic and foreign policies. Nuclear power can serve as insurance, if nothing else. Energy crises make the economy jittery. Because energy demand is rather inelastic, policy makers must ensure a stable supply, including access to oil in foreign countries. Now their difficulties are multiplied by the plausibility of global warming.

Climate change has a global scope, large regional variations, and a long time horizon involving the welfare of future generations. Its impact is fraught with scientific, technological, economic, social, and geopolitical uncertainties. Under such circumstances rational decision makers usually do not rush into a particular course of action, for which doing nothing is a case. They try to keep a portfolio of diverse options, so that they can make successively more definite choices when more knowledge becomes available. Ways to reduce greenhouse gas emissions include cutting energy use, switching to low- or no-emission sources of energy, and sequestering the emitted carbon. For the immediate term, improving efficiency and substituting natural gas for coal may play the largest roles. For longer terms, a slew of technologies are in various stages of development. Some, such as solar and wind power, are more suitable for niche markets than broad use because of their geographical and reliability problems. Others, such as hydrogen fuel and carbon sequestration, are far from commercial viability. Every bit counts. To have substantial impact on global climate, however, a technology must be deployed on a sufficiently large scale. The OTA's argument for preserving the nuclear option becomes more cogent in this context. In 2001 the European Commission's director-general declared that "nuclear power is unavoidable if the European Union hopes to succeed in both maintaining energy supplies and reducing emission levels."[48] That same year the U.S. National Energy Policy recommended "the expansion of nuclear energy in the United States as a major component of our national energy policy."[49]

Nuclear engineers and scientists face great challenges in realizing the expectations of the national policy. They will have to develop new designs for power plants that can be constructed at competitive costs, reactors that are not only safe but also perceived to be safe by the public, fuel cycles that cannot be diverted for weapons proliferation, and long-term spent-fuel repositories with minimal leakage. Reactors in which a core meltdown is impossible have been proposed. One candidate is the pebble-bed modular reactor, whose grain-size uranium fuel kernels are embedded in graphite "pebbles" the size of billiard balls that cannot get hot enough to melt. This reactor is also rather inexpensive to build because of its simple modular design, and inexpensive to operate because of its continuous fueling process. However, it still requires much more research and development.[50]

Hopes of success depend on reinvigorating the human-resource infrastructure, now in sad shape. Half of the nuclear engineering programs at U.S. universities have folded. Faculties are aging and student enrollment has declined. R&D funding has dried up. Although the Energy Department recently launched several nuclear research initiatives, the funding levels are modest. Just as one has to pay a premium to buy insurance, keeping nuclear and other advanced technologies viable as insurance for a steady energy supply requires investment in education and R&D.

Another challenge to nuclear engineers is overcoming adverse social perceptions. Extensive research has revealed what social psychologists call the "social amplification of risk," by which phenomena and technologies that actually pose negligible danger engender excessive fear and mass hysteria. Social amplification is selective; psychologists call those selected technologies whose *perceived* hazard is unreasonably high "stigmatized."[51] Stigmatization contributes to a phenomenon observed by the American Cancer Society: "Public concern about cancer risks in the environment often focuses on unproven risks or on situations in which known carcinogen exposures are at such low levels that risks are negligible."[52] The Cancer Society cites many examples of misguided concerns, among them the low-level radiation near nuclear power plants. This agrees with social psychological research that finds stigmatization extraordinarily severe for nuclear energy, which is much feared although it has never killed or seriously injured anyone in America.[53] Social perception, justified or not, is a political reality that engineers must face. The stigma will stick for a long time, but nothing is forever.

Psychologists find a strong negative association between perceived risk and perceived utility. People tend to regard useful things as less risky. Many deem nuclear power useless because less expensive energy sources are plentiful. This obstacle may diminish as nuclear energy becomes more cost competitive and fossil fuels less attractive for their greenhouse gas emissions. Trust is easy to lose and difficult to regain, but not impossible. Knowledge dampens ungrounded fear and breeds trust, as when people who live near a nuclear power plant tend to become more pronuclear over time. A 2002 Gallup poll on American opinion on energy policy found that 45 percent of respondents favored and 51 percent opposed "expanding use of nuclear energy."[54] This attitude toward nuclear power, although negative, was more sympathetic than that in the 1980s, when support dipped below 30 percent. By consistently maintaining a safe record, improving cost performance, communicating frankly with the public, and working together with all parties, nuclear engineers may regain public confidence and redeem the promissory note of atoms for peace.

7.3 Managing Technological Risks

"If Socrates' suggestion that the 'unexamined life is not worth living' still holds, it is news to most engineers," wrote a philosopher of technology.[55] True ethics is practiced, not merely preached; true meanings of life are lived, not merely talked about. The life to be examined is one's own; for other people's lives, the gossip industry and tabloid journalism are running full steam. In real life, are engineers worse at self-examination than academic philosophers?

Hype and Stereotype

One area that calls for self-examination is exaggerated claims. Hype abounds in the commercial, technical, and literary cultures. Stigmatizing technologies to scare people for publicity is a form of hype. The postmodernist claim that physical laws are thorough social constructions is another; it grossly exaggerates a minor point. Engineering is conducive to hype because many of its results have commercial value. For the same reason, it is subject to harsher discipline. The dot-com bubble in the late 1990s is a spectacular example of technological hype. Its equally spectacular bust shows the ruthlessness of retribution.

An engineering area known for its unending train of burst bubbles is artificial intelligence (AI), where for decades fanciful advertisements for digital computers with superhuman intelligence have succeeded one another. These claims are totally beyond foreseeable capacities of actual computers but are happily exploited and further inflated by journalists and philosophers of the mind. Repeated disappointments harm AI itself by overshadowing its actual achievements, which are substantial. Such hype has been criticized by many, including computer engineers such as Michael Dertouzos, who has exposed many myths regarding information technology.[56]

Another thoughtful engineer worried about the danger of hype was Shannon, who wrote in an editorial: "The subject of information theory has certainly been sold, if not oversold." He exhorted engineers to refrain from loose extrapolations and to focus on research and development on the highest scientific plane. Noting that the transplantation of engineering concepts in unrelated fields and the arbitrary extension of their meanings could generate the false impression of a panacea, he warned: "It will be all too easy for our somewhat artificial prosperity to collapse overnight when it is realized that the use of a few exciting words like *information, entropy, redundancy,* do not solve all problems."[57]

Often when technical concepts are torn out of their mathematically precise context and hyped in some "soft" field, their meanings are so deformed and diluted they become simplistic. Seeing only these crude twists, some constructed by their colleagues, and not bothering to check their originals in engineering and science, many scholars take them as indications of the naïveté of engineers and scientists, adding to the derisive stereotype of the nerd. Thus optimization has been ridiculed as hubris for implying there is a single best way to solve all problems, regardless of the fact that, in engineering, optima are always relative to certain objectives, and that their existence and uniqueness have to be proved. Hype extends technical concepts way beyond their regions of validity. Stereotyping denigrates technical thinking by confusing the hyped concepts for the originals and deriding them. Both are detrimental to the advancement of science and technology.

Are Engineers Ethically Inert?

Stereotypes of engineers as "ethically inert" and engineering as a profession that "has lost its self-conscious moral stance," discussed in section 4.1, are supplemented by more prevalent denouncements of technology, which has been called "the opiate of intellectuals," "a hazardous concept," "the Golem at large," "Frankenstein's problem," "knowledge pregnant with evil." A paper entitled "Ethical Studies about Technology" states: "We say that we have rights to life, but our technologies pollute the water that is necessary for life. We say that we have rights to liberty, but our technologies take away the liberty to breathe clean air."[58] Technology is man-made. Innuendo is prevalent and not always subtle: "Critics decry the lack of consideration for social consequences of technology by those who develop it, identifying engineers as the culprits."[59]

These remarks, easily multiplied, are not from activists' pamphlets but from scholarly books and articles by prestigious professors, many of which are used as reading material for college students. A little examination may show their extreme views to be less criticisms of technology—which are invaluable—than abuses of criticism. Undoubtedly a moral stance is mandatory for engineers; however, engineering students are more likely turned cynical than ethical by a tarring of their future profession. Responsible critiques of technology are crucial but difficult to achieve, for they demand balanced examinations of complex issues. They may be drowned out by indiscriminate condemnations that people easily dismiss as hype for promoting partisan ideologies, like good cars driven out of the market by lemons. Furthermore, exaggerated blaming of technology distracts efforts from other important areas.

For instance, the same ethical study that condemns technology for destroying our rights to life and liberty specifically alleges that "technologies now cause more premature cancers and deaths than any other source, including crime and natural causes." Cancer is, after heart diseases, the biggest killer in America, but to pin those deaths on technology, one must show that it causes most cancers. The ethical study cites as its main evidence a government report finding that only a fraction of cancers are caused by unpreventable internal factors such as genetics. Then it takes as a "fact" that all the remaining cancers are induced by technology, against explicit admonitions of the government report and disregarding detailed data showing that the vast majority of cases are caused by lifestyle factors such as smoking, drinking, and physical inactivity.[60] Let us put aside whether such flexible interpretation in the social construction of "facts" discredits science, ethics, and criticism of technology. Technology does exact tolls that must be addressed, but gross exaggeration of its coercive effect is counterproductive. Besides muddling real problems, it breeds a victim mentality, discourages people from trying to change their lifestyle to reduce risks, and dissuades them from taking responsibility for their own actions. These marginalized considerations are precisely where ethics and humanities education can contribute most.

No one claims technology is omnipotent or omnibenevolent. It is a part of society that must work with many other factors. Engineers cannot do everything; otherwise other experts would be superfluous. For instance, reducing energy consumption, a desirable goal for sustainable development, requires both efficiency and conservation. Efficiency comes mainly from technology, which includes better devices and methods of production. Conservation belongs to consumer behavior, such as paying more up front for energy-efficient appliances and saving later on smaller energy bills. Engineers concentrate on technology and have significantly increased the energy efficiency of everything from industrial plants to home appliances.[61] They are well aware, however, that efficiency alone does not solve the energy problem. Cars now go farther for each gallon of gasoline, but the advantage can be wiped out when consumers switch to bigger sport utility vehicles. Technology and culture have to cooperate for good results. Sociological efforts at putting down engineers' good work as "mere technical fixes" can be better spent on cultivating conservation.

Dealing with Technological Risks

Reliability, safety, environmental benignity, and social acceptability are desirable qualities, but they do not come without a price. Choice is as inevitable in technological designs as in daily actions, and is as much the locus of ethical considerations. To choose is to recognize that resources are finite and afford many uses, and that one cannot have them all. A responsible choice entails awareness of the options abandoned and relinquishment of their benefits and opportunities. It requires trade-offs, in engineers' parlance, or exacts opportunity costs, in economists' terminology. Having to choose is unpleasant and sometimes painful. Ideologues call it crass and evade it by ignoring inconvenient consequences or opportunity costs, such as the immense resources required to alleviate some minuscule risks, resources diverted from other ends that may be more vital. Engineers cannot do that, because they must face realistic constraints and bear the responsibility for their decisions. As marine engineer Edward Wenk wrote: "What makes [engineering] intellectually exhilarating is resolving conflicts between performance requirements and reliability, on the one hand, and the constraints of costs and legally mandated protection of human safety and environment, on the other. Design becomes a tightrope exercise in trade-offs."[62]

The real world is never a safe haven. Life expectancy at birth in

America was only thirty-nine years in 1859. It increased to forty-seven years in 1900 and to more than seventy-six in 1999, partly because of technological advances.[63] Yet the feeling of risk and insecurity also increases, not only because we now have more property to protect and status to secure. Besides danger, risk also connotes agency to reckon with or do something about danger or its consequences. Perils in old days were amply manifested in short life expectancies, and the inability to avoid them may have bred resignation. Science and technology have enhanced our ability to control hazards, made life safer, and opened up immense possibilities, but uncertainties still abound. Furthermore, our endless options increase the difficulty and complexity of decisions. Rational decision making under uncertainty brings the notion of risk to prominence. Economists have calculated the risks of business ventures for a long time. The U.S. Army Corps of Engineers first developed cost-benefit analysis to find the best sites for dams. The 1975 engineering risk analysis of nuclear power plants pioneered the methodology for regulatory policy making.

In engineering and economic analysis, the underlying philosophy is usually utilitarian. Since the writings of nineteenth-century philosophers Jeremy Bentham and John Stuart Mill, *utilitarianism,* which aims to achieve the greatest good for the greatest number of people, has been thoroughly debated. It is a form of *consequentialism,* which evaluates the rightness and wrongness of actions mainly according to the goodness and badness of their consequences. In public ethics and political philosophy, consequentialism often wins over its chief rivals, doctrines that stress individual rights, partly because defining rights beyond the basics is controversial. While rights-based doctrines are criticized for neglecting community values, consequence-based doctrines are criticized for insensitivity to individuals because consequences are often framed in terms of the aggregate good.[64]

In principle, much leeway exists in aggregating. Is distributive justice a good? Is food more good to the starving than to the obese? Should more weight be assigned to the goods of the poor in calculating the aggregate good? If answered in the affirmative, then utilitarianism can include a significant sense of fairness. In practice, however, it is very difficult for utilitarianism to account for distributive justice, partly because not much consensus exists for details. Thus engineering and economic analyses usually leave it aside and assume equal weight for the good of everyone. The rationale is that treating *efficiency* and *equity* in two steps makes complicated problems tractable. Get the biggest pie and then divide it. Efficiency is pursued assuming that the result would be subjected to a compensation mechanism by which the winners pay the losers according to some criteria of fairness. Let us leave aside whether compensation is actually made or whether separating efficiency and equity is a good approximation. The point is that even in principle, utilitarians acknowledge the insufficiency of their calculation for decisions. How safe is safe? How clean is clean? These questions, frequently asked by engineers, are not answered by even the best risk-benefit analysis, although the analysis can provide important clues for decision making.

More problems and uncertainties appear in actual analyses of costs, benefits, and risks. Besides death and injury, what counts as a hazard is itself problematic, not to mention estimating its probability. Three classes are especially difficult. The first is where hazards interact. The second includes extremely rare hazards, such as accidents in nuclear power plants. The third includes exposure to very low doses of toxins; the etiology of pathology is obscure and extrapolations of models and data are controversial. Estimations for these risks are only crude approximations or educated guesses. Furthermore, calculating costs and benefits for plant safety, environmental impact, and similar situations sometimes requires putting price tags on life, lifestyles, welfare of future generations, and various aspects of nature. All these estimates are fraught with value judgments, for which analysts have to secure some consensus from relevant parties through proper procedures.[65]

Consequentialism is anything but perfect, but critics seem unable to offer *practical* alternatives that go beyond sound-good rhetoric. To get things done, utilitarian analyses are valuable when they are recognized as instruments of decision making and not the ultimate decision makers. They open conceptual frameworks where assumptions can be laid out, uncertainties exposed, values discussed, evidence presented, results criticized, and improvements suggested. By applying criteria of risk assessment consistently, they allow people to compare the relative probabilities of various risks, although the probability value of a single risk is only a crude estimate. By making relevant factors explicable to the public and providing justificatory reasons to decision makers, they help to legitimize choices. Widely used in the government since the 1970s, cost-benefit analyses of proposed legislation lessen the pressure of special interest groups. The Supreme Court faulted OSHA for failing to perform a risk-benefit analysis before demanding a tenfold reduction in the level of benzene to which workers are exposed. OSHA later performed the required analysis, which supported its stringent standard. Although the decision stayed the same, the relevant assumptions underlying it are now more transparent and hence more socially acceptable.

Unceasing Vigilance in Safety Engineering

The joke goes that Microsoft's president boasts that if cars were like computers, a Cadillac would now cost only a hundred dollars. Whereupon GM's president retorts that if cars were like computers, they would crash mysteriously every day or two. The erratic performance of Microsoft's Windows, frowned on by computer professionals, is tolerated by consumers because of its relatively low price and the minor consequences of its failure. A computer crash kills not people but only work; consumers sigh resignedly and reboot. Reliability—freedom from breakdowns and minor problems—is a mark of high quality and an important criterion in engineering. Generally, reliability increases as a technology matures. The plain old telephone is among the most reliable of services, with availability exceeding 99.999 percent, which implies an average of less than 315 seconds of down time in a year. The newer technologies of wireless mobile and packet-switched Internet communications are less reliable, but they have a lofty goal to aspire to.

For brevity, I lump under "safety" the prevention of those rare failures that have disastrous consequences, which include extensive casualties and property damage. Such failures are rare because a technology that frequently caused unintended catastrophes would soon be abandoned. Safety usually improves as engineers learn from mistakes. Steam boilers for anything from home heaters to industrial turbines can be lethal. A boiler explosion on a Mississippi riverboat killed 1,450 Union soldiers. Boilers routinely blew up in the nineteenth century, reaching a peak of more than 400 cases annually in America by 1900. Then the tide was turned by the ASME's Boiler Code, which has been adopted as law by many states. Now boiler explosions are rare.

Engineers can make a system extremely safe but not absolutely so. Surprise is always a possibility. Often many technical, human, and social factors, perhaps triggering each other, contribute to an accident. For example, in the decision process that ended in the explosion of the space shuttle *Challenger,* politics trumped engineering, epitomized when a senior manager told the engineer arguing for canceling the launch to "take off his engineering hat and put on his management hat."[66] Even in this paradigmatic case for the social construction of disaster, however, engineers cannot wash their hands. A flawed O-ring design lay at the bottom of the mishap. No matter how much the situation was compounded by mismanagement, the design fault rests on engineers. Avoiding designs that can lead to disasters through human error or natural accident demands proper integration of physical and organizational technologies in designing and operating technological systems, especially complex and high-stakes ones.

The Code of Hammurabi decreed in the third millennium B.C.E.: "If a builder builds a house and the house collapses and causes the death of the owner, that builder shall be put to death." Penalties are less harsh nowadays, but responsibilities are no less onerous. Hoover wrote: "The great liability of the engineer compared to men of other professions is that his works are out in the open where all can see them. His acts, step by step, are in hard substance . . . If his works do not work, he is damned. That is the phantasmagoria that haunts his nights and dogs his days."[67]

In designing complex systems, engineers introduce redundancy, cross check everything, and when in doubt prefer to err on the safe side and use tried methods, not because they are stodgy, as some cultural stylists claim, but because they know that their products affect the lives of many people. The required vigilance is unrelenting. Joseph Gavin recalled in describing the design process for the Apollo Lunar Module that put the first human on the moon: "Those things we worried about didn't give us a problem. The places we didn't worry about is where we ran into problems." What they had worried about operated smoothly because they had anticipated and ironed out the kinks. Most worrisome was the possibility of oversight, and engineers kept their fingers crossed despite their best efforts. The experience was so nerve-wracking that some were relieved to see the Apollo project end. "You can't keep throwing sevens all the time," remarked Robert Gilruth, who headed the Johnson Manned Spacecraft Center.[68]

Here experience can become a double-edged sword. Experienced workers are efficient because they know what works, often without having to focus on the details, but that efficiency can also let problems slip by. To minimize the chance of overlooking something, Augustine urges engineers to "question everything." Ask not only experts and workers involved in the project but also outsiders: someone with no prior experience, someone who can say the emperor has no clothes.[69] Two cases illustrate the difference a vigilant and open attitude can make.

NASA had ample reason to contract Perkin-Elmer (now Hughes-Danbury) to manufacture the primary mirror for the Hubble space telescope in 1981. A leader in optical instruments, the company was also seasoned in making mirrors for military spy satellites. Nevertheless, all its experience did not prevent it from polishing the Hubble mirror into the wrong shape. Perkin-Elmer employed the newest technology, using an ultraprecise device to test the mirror all through the process of polishing. The device caught and corrected the finest deviations. Unfortunately, it got misaligned when it was transported from the military to the civilian side of the company, so that it was blind to a gross error. The error was in fact detected by a quick and dirty test, but the test result was dismissed as crude and implausible—no one so experienced could make such a stupid mistake. But Perkin-Elmer did. To prevent similar mistakes from happening again, the board of inquiry into the failure demanded an organization "which encourages concerns to be expressed and which ensures that those concerns which deal with a potential risk to the mission cannot be disposed of without appropriate reviews."[70]

A Near Miss at Citicorp Center

Alertness to innocent questions saved the Citicorp Center from a disaster that might have dwarfed the Hyatt-Regency walkway collapse that killed 114 and injured more than 200. Soon after the Citicorp Center in New York City was occupied in 1977, its chief structural engineer, William LeMessurier, was asked by a student about the peculiar position of its pillars and was prompted by the conversation to use the case in a course he taught at Harvard. A tall building's structure must bear a vertical gravity load and a horizontal wind load with adequate strength and stiffness. For a wind brace, the Citicorp Center has a steel frame concealed behind its curtain wall, as illustrated in Fig. 7.2a. In preparing a variety of problems for his class, LeMessurier calculated the force exerted by a forty-five-degree wind and found that it increased the strain on the steel braces by 40 percent. His original design had enough safety margin to take the increase in stride. However, he had discovered a month earlier that his associates had changed the design from welded joints to bolted joints, arguing that the former were expensive and unnecessary. Design changes are common practice. Bolted joints conforming to the industry code for columns would do fine.[71]

Putting the matter aside would have been easy for a man as busy as LeMessurier. A year later, the structural engineers for the Hyatt-Regency Hotel in Kansas City, Missouri, would leave alone a minor design change for its walkways (Fig. 7.2b). Designs and changes in both cases involved associates, contractors, subcontractors, and communication among many people in many organizations. Nevertheless, responsibility for final approval belonged to the chief engineers, whose vigilance even in checking superficially reasonable changes can be a knife edge between safety and catastrophe. The engineers for Hyatt would be convicted of criminal negligence and professional misconduct after the change they overlooked caused the hotel's walkways to collapse. The engineer for Citicorp decided to double check.

LeMessurier found that his associates had treated the braces as trusses instead of columns and used perilously few bolts. Detailed calculations and wind-tunnel tests revealed more problems, which he summarized in a document entitled: "Project SERENE (Special Engineering Review of Events Nobody Envisioned)." The consequence of the unforeseen events was the risk that the building would collapse in winds of seventy miles per hour, which according to meteorological data had a one in sixteen chance of occurring in each year. August had just begun; the hurricane season was young.

Agonizing over the prospect of a ruined career, the engineer notified the architect. Together they informed Citicorp that its new headquarters was structurally flawed but could be fixed by welding steel plates over each of the building's more than two hundred bolt joints. Citi-

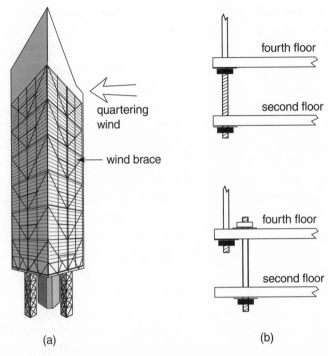

(a) (b)

Figure 7.2 (a) The fifty-nine-story Citicorp Center sits on nine-story-high stilts: a central core and four pillars located under the centers of its sides to leave the corners hanging. Under one corner is tucked St. Peter's Church, which owns the corner lot but agreed to exchange its air rights for a new building that had to be completely freestanding. The building has a braced frame structure that, besides sharing the gravity load, carries most of the lateral wind load. The discovery that a quartering wind striking two surfaces simultaneously would place an unexpectedly large strain on its joints led to an emergency repair in 1978. (b) In the original design for the Hyatt-Regency Hotel, the fourth- and second-floor walkways were supported by box beams hung on common rods from the atrium ceiling. As built, the second-floor beams hung on the fourth-floor beams by separate rods. The altered design saved on threading long portions of the common rods but doubled the load on the fourth-floor beams, leading to the eventual collapse. *Source:* (a) Adapted from B. S. Smith and A. Coull, *Tall Building Structures* (New York: Wiley, 1991), p. 127; (b) see A. Jenney (1998), www.uoguelph.ca/~ajenney/webpage.htm.

corp's top management was reasonable. Competent professionals all, they swiftly decided on a course of action and secured the cooperation of municipal officials and industrial contractors. The building's entire structure was reevaluated, and the immediate production of special steel plates was arranged. Welders from several states were assembled, their torches lighted up the night, sparing the offices for bankers during the day. The press was given sufficient information without creating public hysteria. The mayor's office and the Red Cross quietly drew up plans for emergency evacuation and disaster. Luckily the plans never had to be activated, although Hurricane Ella brought several anxious days. Repairs were finished in October, leaving the building strong and sound.

Citicorp did not sue. For costs of the emergency repair, it collected $2 million, the maximum covered by LeMessurier's liability insurance, and swallowed the balance, rumored to be in the millions. The insurance company paid but did not raise LeMessurier's premium. A general feeling seemed to be voiced by Citicorp's chief contractor: "It wasn't a case of 'We caught you, you skunk.' It started with a guy who stood up and said, 'I got a problem, I made the problem, let's fix the problem.' If you're gonna kill a guy like LeMessurier, why should anybody ever talk?"[72]

"To engineer is human," Petroski aptly entitled his book on engineering failures. The list of engineering disasters is long: *Challenger* and *Columbia,* Three Mile Island and Chernobyl, collapse of the Tacoma Narrows Bridge and other structures, toxin-spewing fires at chemical plants in Bhopol and elsewhere, airplane crashes caused by various technical problems. The list of near misses, such as that at Citicorp, is much longer. Well aware of human frailty, engineers do their utmost not only to prevent accidents but also to learn from disasters and near disasters, so that they do not repeat mistakes. In response to the Hyatt disaster, the American Society of Civil Engineers took strong actions and now holds engineers responsible for all aspects of structural safety in their building designs. Petroski wrote that the concept of failure "is central to understanding engineering, for engineering design has as its first and foremost objective the obviation of failure. Thus the colossal disasters that do occur are ultimately failures of design, but the lessons learned from those disasters can do more to advance engineering knowledge than all the successful machines and structures in the world." "For this reason it is important that engineers study failures at least as much as, if not more than successes, and it is important that the causes of structural failures be as openly discussed as can be."[73]

Engineering for the Environment

Remains of Roman aqueducts testify to one of the first vital jobs of engineering: to supply fresh water to population centers. Ideologues who accuse technology of destroying our right to life by polluting our water have forgotten about its role in cleaning up perhaps the most lethal pollution, germ-breeding human and animal wastes. Soon after the 1854 cholera outbreak in London, during which epidemiologist Edwin Chadwick proved that the disease was spread by contaminated water, what was called *sanitary* or *public health engineering* emerged to take on the tasks of supplying clean drinking water and treating wastewater. Besides building sewers and other physical infrastructures, engineers used rapidly developing knowledge of vector-borne infectious diseases to filter and disinfect water, so that by World War II, drinking water from the tap became the norm in the United States. They also developed sedimentation, anaerobic digestion, and other chemical and biological processes to treat wastewater, not only to eliminate water-borne diseases but also to prevent harm to the environment. Now called environmental engineers, they take care of all public issues concerning water.[74]

Technology cleans up some pollutants and introduces others. Metal mining, chemical processing, electricity generation, and other industrial activities release toxins from localized sources. Automobile emissions and agricultural runoff are dispersed sources of pollution. Local efforts for environmental protection, already under way in the nineteenth century, became a massive societywide movement in the 1960s. Responding to society's call, environmental engineering has expanded to cover air, water, and land. Engineers team up with natural scientists to understand acid rain, ozone depletion, and global warming, and to find ways to eliminate the causes and minimize the effects. To maintain air and water quality, new technologies must be developed to control or prevent pollution by a lengthening list of chemicals. To manage hazardous wastes, processing and repository sites must be prepared. These tasks call for diverse expertise. The American Academy of Environmental Engineers enjoys sponsorship from societies of civil, mechanical, and chemical engineers and other scientific organizations.

Now some 52,000 environmental engineers work in the United States. They design infrastructures for public health. Some participate in drafting environmental regulations; others work in the environmental industry that has sprung up to prevent or monitor pollution, manage wastes, clean up contaminated sites, and produce the requisite equipment. Of this industry's revenue, which exceeded $200 billion in 2000, about 9 percent goes to engineering and consulting, testing, and technologies for processing and prevention.[75] In addition, environmental considerations penetrate all relevant engineering projects. For instance, technologies for mitigating environmental impact are among the top priorities in mining and petroleum engineering.

In the last four decades of the twentieth century, the number of environmental laws and regulations increased more than sixfold in the United States. At first regulations adopted a command attitude, and engineering and industry responded by meeting demands one at a time. Regulators set the standard for automobile emissions, for example, and engineers responded by developing catalytic converters. Industrialists had no incentive to support more innovations that might invite even stricter standards. Attitudes began to change in the late 1980s. Slowly, regulators shift their emphasis from pollution abatement to prevention, from command to economic incentive. Business leaders realize that environmental consciousness can be profitable. Engineers change from being responsive to being proactive, jumping ahead to develop green technology.[76]

Most old environmental measures are end-of-the-pipe treatments, such as smokestack scrubbers, which capture pollutants after they are produced. Such technologies are still useful, but engineering's scope has broadened. Systems engineering, which takes a comprehensive view of product and production, is already germane to sustainable development.[77] Green engineering extends the view to cover the whole supply chain and all side effects of a production process.[78] Its *life-cycle analysis* assesses the environmental risks of various products by comparing their entire manufacturing process, tracking their components and byproducts from extraction of raw materials to disposal of wastes and de-commissioned products. Taking into account the pollution generated in the manufacture of efficient solar panels, for instance, solar energy may not be as clean as many environmentalists think.

Green engineering aims to design environmental benignity into a product, so that it can be produced with minimal energy and raw materials, few hazardous steps, and little if any harmful by-product and wastes that must be disposed of. Recycling, reuse, and other environmental factors are taken into consideration from the very beginning of design. Industrial plants are treated as integral elements in the natural ecosystem. These are difficult goals, and their attainment requires not only tremendous research and development but also full cooperation with industry and environmentalists. Green technology is still in its infancy, but many programs are under way, such as those to find environmentally friendly ways of manufacturing semiconductors and computer chips.[79]

Far from being a profession that has lost its moral stance, engineering is raising its moral consciousness. It is taking on more social responsibility through life-cycle analysis, green engineering, systems engineering, and other increasingly widespread practices. Many new engineering textbooks begin with a chapter on social context, including the harmful environmental impacts of the very technologies covered in the book.

Engineers are never in a position to be complacent. Soaring technological complexity and risk keep them forever on the edge. After the *Challenger* disaster, engineers who blew the whistle on NASA and its contractors went on to push for ethical education. Many engineering schools responded by introducing courses on professional ethics. Vannevar Bush saw a clear role for the engineer in his vision of the endless frontier: "The impact of science is making a new world, and the engineer is in the forefront of the remaking . . . He builds great cities, and builds also the means whereby they may be destroyed. Certainly there was never a profession that more truly needed the professional spirit, if the welfare of man is to be preserved."[80]

Statistical Profiles of Engineers

"Counting the S&E Workforce: It's Not That Easy" is the title of a report by the National Science Foundation (NSF). Education is an easy criterion for counting; fields of specialization are printed on college diplomas. However, the NSF cautions that most people do not work in the field in which they were educated. Many people with science and engineering degrees work in areas such as management and high-school teaching.

If counting S&E professionals is not easy, distinguishing between scientists and engineers is even more difficult. Often they bear the same title in industrial R&D facilities, as with the technical staff at Bell Laboratories, because they perform similar work. Even in universities with separate science and engineering schools, the intellectual content is often intermixed. Members of science departments publish in engineering journals and vice versa. The magazine of the American Chemical Society is titled *Chemical and Engineering News*, and 15 percent of the members of the American Physical Society are engineers. This overlap is increasing as scientists become more practical and engineers perform more basic research. Most census and surveys reflect respondents' self-identification, but as with all polls, answers depend on what categories are offered and how questions are phrased. Consequently, the counting results vary dramatically according to the criteria used to define an engineer or scientist.

The U.S. Census Bureau's 2000 tally found 2.1 million working engineers who, together with 2.1 million computer professionals, made up 3.1

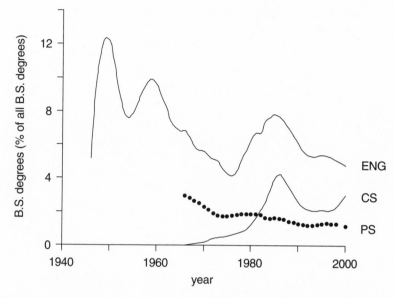

Figure A.1 Annual bachelor of science degrees awarded in engineering (ENG), computer science (CS), and the physical sciences (PS) as percentage of the degrees granted in all fields. *Source:* National Research Council, *Engineering Education and Practice in the U.S.* (Washington, D.C.: National Academy Press, 1986), p. 89. National Science Foundations, *S&E Degrees*, tables. 1, 26, 35, 46.

percent of America's civilian workforce of 135 million. The engineers' share of the workforce has remained almost constant since 1970, after tripling in the preceding thirty years.[1] The NSF's numbers for S&E personnel are much lower than those of the Census Bureau or the Bureau of Labor Statistics. While the bureaus count many technicians and computer programmers, according to the NSF's criteria almost all working scientists and engineers are holders of university degrees. I use mostly NSF results.

Engineers in School

Of all freshmen students in 1946, more than 14 percent enrolled in engineering, which seemed to be a favorite among veterans. Soon the number of baccalaureates in engineering overtook those in the physical sciences. The number of doctorates took longer to build up, but it did grow (Figs. A.1, 2.1).

Table A.1 Degrees in various fields granted by U.S. universities in 2000, percentage of masters and doctorates among degrees granted, and degrees in a field as percentage of the degrees in all fields for the same level.

Field	Degrees in the field			Percentage of all fields		
	Total (000)	% M.S.	% Ph.D.	B.S.	M.S.	Ph.D
Engineering	90.5	28	6	4.75	5.63	12.8
Aeronautics	2.1	28	11	0.10	0.12	0.9
Chemical	8.3	17	9	0.50	0.30	1.7
Civil	14.3	29	4	0.77	0.91	1.3
Electrical	27.6	30	6	1.41	1.83	3.7
Industrial	7.2	43	3	0.31	0.67	0.4
Mechanical	17.4	19	5	1.05	0.74	2.1
Material	2.2	35	21	0.08	0.17	1.0
Computer sci.	52.8	28	2	2.98	3.18	2.0
Mathematics	16.1	21	7	0.94	0.72	2.9
Physical sci.	21.5	16	16	1.17	0.77	8.2
Life sci.	100.1	10	7	6.63	2.24	16.4
Social sci.	141.1	17	3	9.06	5.13	10.1

Source: National Science Foundation, *S&E Degrees.*

At the beginning of the twenty-first century, American universities annually award about 97,000 baccalaureate and 6,200 doctoral degrees in engineering and computer science. They amount to about 8 percent of all bachelor of science degrees, 9 percent of all master's degrees, and 15 percent of doctor's degrees in all fields (Table A.1).

About one half of engineering college graduates go on to graduate school. The ratio of master's to bachelor's degrees awarded annually is about 40 percent. A study conducted by the American Society for Engineering Education in 1968 recommended a specialized five-year master's degree to be the basic program for professional engineers. That recommendation has been controversial, for engineering is not homogeneous and the work content of various branches of engineering differ.

In number, more doctorates in engineering than in the physical sciences graduate from American universities every year. In the ratio of advancement, a smaller percentage of engineers continue to the doctorate level. In engineering, the number of doctorates awarded is about one sixteenth that of baccalaureates; in the physical sciences, the ratio is about one fourth. If we regard the *number* of doctoral degrees as an indication of the *amount* of research activity, we see that engineering has more research opportuni-

Table A.2 Occupational distribution of employed U.S. scientists and engineers in 1999, by fields in which the bachelor of science is the highest degree held.

B.S. degree in	Percentage employed in					
	Eng.	Comp.	Phys.	Life	Social	Non-S&E
Engineering	53.8	9.3	0.6	0.2	0.1	36.0
Computer sci.	2.1	58.5	—	—	—	39.3
Physical sci.	10.5	7.3	24.3	2.4	0.3	55.3
Life sci.	1.7	2.9	2.9	10.9	0.2	81.4
Social sci.	0.6	4.5	0.3	0.3	2.6	91.8

Source: National Science Board, S&E Indicators 2002, table 3-6.

ties than the physical sciences, but computer science has less than mathematics. If we regard the *ratio* of Ph.D.'s to B.S.'s in a field as an indication of the field's research *intensity*, then engineering is comparable to mathematics, both being less research-intensive than the physical sciences but more than computer science. Research seems to be the major occupation for physical scientists but not for engineers.

In all fields, those who have gone through graduate school are more likely to work in the field of their degree. The work pattern differs widely for those who stop at a bachelor's degree, as shown in Table A.2. A much larger fraction of engineering and computer science graduates work in their field of education than natural science graduates. Undergraduate students in engineering and computer science are thus more likely to find what they are learning to be directly useful in their future occupation than their counterparts in other fields.

Comparing the ratio of advanced degrees to bachelor's degrees in a field with the proportion of degree holders who work in the field, one can surmise that without a doctoral education, there is not much employment opportunity in the physical sciences. Engineers have more choices because of their practicality. More than half of those who stopped with a bachelor's degree in engineering work as engineers. Fresh graduates with engineering degrees are likely to get higher salaries than their counterparts with science degrees, sometimes significantly higher. Fewer engineers tend to go for a Ph.D., partly because the opportunity cost for such a lengthy education is higher in engineering than in many of the sciences. Experience on the job is also more valuable in engineering than in other fields.

Engineering graduates say they are more satisfied with their choices. Af-

Table A.3 Characteristics of U.S. engineering and science workforce in 1999: total number of employed degreed engineers and scientists; percentage holding masters and doctorate degrees; percentage workers with primary or secondary activities in R&D; median annual salaries (in thousands of dollars); percentage employed in industry, academia, and government.

Field	Total (000)	% M.S.	% Ph.D.	% in R&D		Salary (000)		% in sector		
				B.S.	Ph.D.	B.S.	Ph.D.	.com	.edu	.gov
Engineering total	**1,370**	**28**	**6**	**45**	76	**60**	79	**81**	**5**	**14**
Aerospace	68	39	7	41	85	69	84	76	4	21
Chemical	80	26	10	51	75	65	80	94	3	4
Civil	224	26	2	36	67	55	70	64	2	35
Electrical	362	30	5	49	76	65	86	86	3	11
Industrial	82	21	1	30	62	55	85	93	2	5
Mechanical	266	23	3	55	80	60	75	93	2	6
Computer/information	**1,058**	**29**	**3**	**34**	**72**	**61**	**81**	**88**	**5**	**8**
Mathematics	36	44	22	21	67	56	74	48	22	31
Physical sciences	298	25	29	37	73	45	70	54	28	18
Life sciences	342	21	35	23	68	37	62	33	48	19
Social & human sciences	363	43	35	13	46	30	60	43	45	12

Source: National Science Board, *S&E Indicators 2002*, tables 3-10, 3-12, 3-22. Percentages in R&D are for 1997 from *S&E Indicators 2000*, table 3-27.

ter working for five years, fewer than 10 percent of electrical engineers say they would probably choose another field if they could start all over again, compared with 18 percent of biologists and 24 percent of physicists.[2]

Engineers at Work

Engineers are mostly male and are ethnically diverse. In 1999, 19 percent of engineers and computer professionals in the United States were ethnic minorities, compared with 16 percent of life and physical scientists, 13 percent of social and human scientists, and 17 percent of the U.S. workforce as a whole.[3]

In gender balance, America lags behind some other countries. Only 10 percent of engineers and 27 percent of computer professionals are female. Less than 20 percent of the undergraduate engineering enrollment is female, which is comparable to physics but lower than in chemistry and the life sciences. Female enrollment in computer science peaked at 37 percent in the mid-1980s but now falls below 30 percent, which is way below that in mathematics.[4]

The characteristics of university-degree-holding scientists and engineers in the workforce can be seen in Table A.3. Here the differences between engineers and scientists show up. More scientists work in academia and have more formal education; those without graduate schooling tend to drop out of scientific occupations. Engineers and computer professionals have a broader distribution in their level of formal education because more baccalaureates stay in the field.

About 80 percent of workers who identify themselves as engineers work in the private sector. About one half have research and development as their primary or secondary activity. Management is the next most common activity, followed by education, production, and inspection. In many large technology-oriented companies, several levels of organization exist between "pure engineers" and "pure managers." Positions in these ambiguous middle levels, which involve both kinds of work, are often filled by engineers, for they are more likely to have the technical ability to gain the trust of the engineers whom they supervise. The large proportion of managerial jobs held by engineers highlights the importance of technical knowledge in organizing the productive force for a technological society. This also explains the high percentage of engineers in sales and marketing. Technical expertise is necessary for explaining the function and operation of high-tech products to potential customers.

U.S. Research and Development

Partly because of the public mistrust of government meddling, for a long time the U.S. government kept out of science and technology, except for scattered mission agencies, such as the Geological Survey for mapping the country and the Census Bureau for statistical analysis. The Army Corps of Engineers, established in 1802, has operated many laboratories in its history. Among the first was the Waterways Experimental Station at Vicksburg, Mississippi, founded in 1929 with the aim of understanding complex river dynamics for effective flood control.[1] Another early mission agency, founded in 1915, was the National Advisory Commission on Aeronautics (NACA), precursor to NASA. Its Langley Aeronautical Laboratory played a great role in the development of aeronautical engineering. When few universities had degree programs in aeronautics, Langley recruited graduates from mechanical and electrical engineering and trained them in aviation-related R&D. Before it turned to concentrate on military aircraft, it was a vital R&D center that led to the modern passenger airliner.[2]

The first attempts at orchestrating technological efforts on a national level occurred during the two world wars, especially World War II. Impressed by the practical potential of scientific knowledge previously deemed purely academic, the U.S. government entered R&D with a big purse and assumed some coordinating functions. Congress lavished R&D funds on the mission-oriented agencies for defense, energy, and health. The National Science Foundation, created in 1950, included engineering from

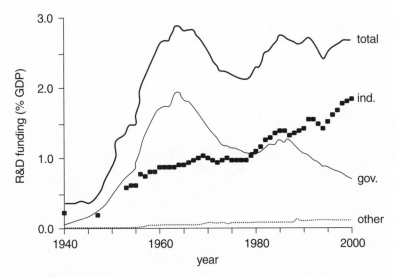

Figure B.1 R&D expenditures in the United States as percentage of GDP: total, industry, federal government, and nonprofit organizations. *Source:* National Science Board, *Science and Engineering Indicators 2002,* tables 4.1, 4.6, 4.10.

its inception and created an independent engineering directorate in 1981, leaving no ambiguity about the equal footing of science and engineering.

Fueled by the cold war, the space race, economic prosperity, new high-tech industries, and global competition, America's R&D expenditures rose from $7.4 billion in 1947 to $249 billion in 2000 (both in 1996 dollars). The steadily rising dollar amount obscures the fact that the national economy has also been growing. Important quantities are dimensionless, as engineers say. A dimensionless measure here is R&D *intensity,* or R&D spending as a fraction of the nation's gross domestic product (GDP). As shown in Fig. B.1, intensity peaked in 1964, when R&D absorbed 2.9 percent of America's GDP. Since 1975, the United States' R&D intensity has been matched by Germany and Japan. These R&D expenditures indicate the importance of science and technology for economic competitiveness.

More than one half of federal R&D funding goes to development, of which engineering has a lion's share. Engineering's share in research is smaller but is still significant, as shown in Fig. B.2. It decreased in recent years because of the relative decline in the defense and space programs. Cuts occurred mainly in applied research. Over the decades, engineering has steadily received about 10 percent of federal funding in *basic* research, and this does not include the funding that goes to computer science.

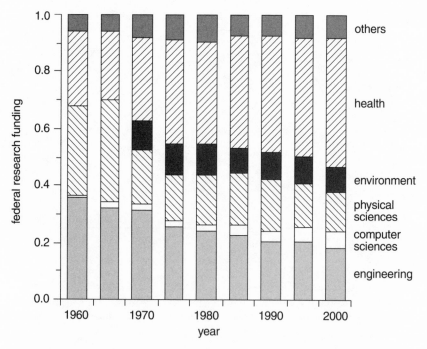

Figure B.2 Distribution of federal obligations for research (basic and applied) among science and engineering subjects. *Source:* National Science Foundation, *Survey of Federal Funds for R&D,* various years.

Since the end of the cold war, the bills for R&D and even basic research increasingly are being picked up by the private sector. Two measures of an industry's technological *intensity* are its R&D spending with respect to its net sales and the ratio of R&D staff with respect to total employees. In Table B.1 we see that the aerospace industry is highest in technological intensity, partly because of the extreme performance demanded by military aircraft and missiles. Its R&D expenditures have dropped drastically since the end of the cold war.

More data and graphs are provided at www.creatingtechnology.org.

Table B.1 1998 industrial R&D spending in billions of current dollars and as percentage of net sales; total number of R&D scientists and engineers and number per 1,000 employees in R&D-performing companies.

Industry	R&D expenditures		R&D personnel	
	Bill. $	% Sales	Total (000)	No. per 1,000
Manufacturing	120.4	3.7	659.1	5.7
Aerospace	14.5	9.3	66.4	9.1
Automotive	14.8	—	62.8	4.9
Chemical	21.8	6.5	90.7	9.7
Drug and medicine	12.6	10.6	50.0	15.7
Electrical/electronic	26.0	7.1	173.8	12.5
Scientific instrument	14.1	—	58.4	9.0
Machinery	14.9	5.1	96.1	7.9
Service	29.3	11.0	196.7	11.1
Engineering/management	11.8	19.3	57.6	15.1
Business	15.2	10.7	127.5	12.0
Health	1.2	8.6	4.6	5.7
All industries	169.2	3.6	997.7	5.5

Source: National Science Foundation, Research and Development in Industry 1998, NSF 01–305.

1. INTRODUCTION

1. Sloan quoted in C. C. Furnas and J. McCarthy, *The Engineer* (New York: Time Life Books, 1966), p. 9; H. Hoover, *The Memories of Herbert Hoover: Years of Adventure, 1874–1920* (New York: Macmillan, 1951), p. 132. The word *he* has three common usages nowadays. The first two are time-honored dictionary definitions used in such places as the U.S. Constitution: first as a pronoun for a particular male, and second as a gender-blind pronoun for an arbitrary person. The third usage, of recent vintage, is gender conscious and discriminatory, as in, "The engineer likes his work (I use *his* deliberately because engineers are mostly male)," which snubs the minority. I use and understand all usages of *he* and its cognates only in the first two senses, unless an author explicitly states otherwise. A similar attitude applies to the word *man.*

2. V. Bush, in *Listen to Leaders in Engineering,* ed. A. Love and J. S. Childers (Atlanta: Tupper and Love, 1965), pp. 1–15. For its fiftieth birthday in 2000, the NSF posted Bush's *Science—The Endless Frontier* on its website: www.nsf.gov/od/lpa/nsf50/vbush1945.htm.

3. J. C. Maxwell, *A Treatise on Electricity and Magnetism* (New York: Dover, 1873), p. vi.

4. W. G. Vincenti, *What Engineers Know and How They Know It* (Baltimore: Johns Hopkins, 1990), p. 6.

5. N. R. Augustine in *The Bridge* 24(3): 3 (1994).

6. Clarke quoted in *The Profession of a Civil Engineer,* ed. D. Campbell-Allen and E. H. Davis (Sydney: Sydney University Press, 1979), p. 204.

2. TECHNOLOGY TAKES OFF

1. H. Collingwood, *Harvard Business Review* 79 (11): 8 (2001).
2. Socrates, *Apology,* 21. For a brief exposition of Greek terms, see F. E. Peters, *Greek Philosophical Terms* (New York: New York University Press, 1967), and A. Edel, *Aristotle and His Philosophy* (Chapel Hill: The University of North Carolina Press, 1982).
3. Plato, *Gorgias,* 465, 500.
4. Aristotle discussed the rational faculties in *Ethics* 1139–1141, to which he referred in *Metaphysics* 981 and 1025, where he elaborated on the characteristics of art. The analysis of cause occurs in *Metaphysics* 1013–1014 and *Physics* 194–195. J. L. Ackrill, ed., *A New Aristotle Reader* (Princeton: Princeton University Press, 1987), pp. 419, 255–256.
5. A. Grafton, *Leon Battista Alberti* (New York: Hill and Wang, 2000), p. 80.
6. J. Bigelow, *Elements of Technology* (Boston: Boston Press, 1829), pp. iii–iv.
7. D. Hill, *A History of Engineering in Classical and Medieval Times* (La Salle: Open Court, 1984).
8. J. K. Finch, *Engineering Classics* (Kensington, Md.: Cedar Press, 1978), pp. 7–8. H. Straub, *A History of Civil Engineering* (Cambridge, Mass.: MIT Press, 1952), pp. 31–32.
9. B. Gille, *Engineers of the Renaissance* (Cambridge, Mass.: MIT Press, 1966). Grafton, *Alberti,* chap. 3.
10. Grafton, *Alberti,* p. 77.
11. In 1481, at the age of twenty-nine, Da Vinci wrote a letter in which he listed his own qualifications, starting with "I can construct bridges very light and strong, and capable of easy transportation" and going on and on in military and civilian engineering before ending with architecture and sculpture. Gille, *Engineers,* pp. 125–126.
12. W. H. G. Armytage, *A Social History of Engineering* (London: Faber and Faber, 1976), p. 45. J. B. Rae and R. Volti, *Engineer in History* (New York: Peter Lang, 1993), pp. 86–87.
13. For instance, "*Applied science* is often regarded as a synonym for 'engineering'" is asserted on p. 6 of *What Can Be Automated?* ed. B. W. Arden (Cambridge, Mass.: MIT Press, 1980), a carefully edited report with contributions from eighty engineers and scientists from academia, industry, and the government.
14. R. K. Merton, *Social Theory and Social Structure* (New York: Free Press, 1968), p. 663. Not only Isaac Newton but "every English scientist of this time [seventeenth century] who was of sufficient distinction to merit mention in general histories of science at one point or another explicitly re-

lated at least some of his scientific research to immediate practical problems."

15. The attack on engineering as applied science, probably initiated by E. T. Layton, Jr., *Technology and Culture* 17: 688 (1976), and now widespread in technology studies, is based on a peculiar notion of applied science as "debased," "more or less mechanical," and "introducing no new knowledge." This groundless notion badly caricatures applied science and science in general. A refutation is given at www.creatingtechnology .org/applsci.htm.

16. The National Science Foundation regularly surveys Americans' attitudes toward science and technology. In its 2001 survey, 86 percent of respondents said they thought that science and technology were making lives healthier, easier, and more comfortable; 85 percent thought they would bring more opportunities for the next generation; 72 percent thought their applications make work more interesting (*Science and Engineering Indicators 2002*, Tables 7–12).

17. G. Galilei, *Dialogues concerning Two New Sciences* (New York: Dover, 1954), p. 1.

18. Galelei, *Dialogues*, p. 2. This is repeated on p. 112, where he analyzes the cantilever beam.

19. For a history of materials study, see T. F. Peters, *Transitions in Engineering* (Basel: Birkhäuser Verlag, 1987) and Straub, *Civil Engineering*, pp. 107–110. Finch, *Engineering Classics*, describes classic literature.

20. Peters, *Transitions*, p. 55.

21. Straub, *Civil Engineering*, p. 116.

22. E. Benvenuto, *An Introduction to the History of Structural Mechanics* (New York: Springer-Verlag, 1991), chap. 1; pp. 351–358.

23. Peters, *Transitions*, p. 51. Coulomb quoted in Benvenuto, *Structural Mechanics*, p. 386.

24. Peters, *Transitions*, p. 53. Navier is also discussed in Straub, *Civil Engineering*, pp. 152–158.

25. N. Rosenberg and W. G. Vincenti, *The Britannia Bridge* (Cambridge, Mass.: MIT Press, 1978). T. F. Peters, *Building in the Nineteenth Century* (Cambridge, Mass.: MIT Press, 1996), pp. 159–178.

26. Census Bureau, *Economic Census 1997*, www.census.gov/epcd/ec97. The remaining 40 percent of the construction industry belongs to special trade contractors for electrical, plumbing, heating, and other systems.

27. R. F. Jordan, *A Concise History of Western Architecture* (London: Harcourt, Brace, Jovanovich, 1969), p. 295.

28. H. R. Hitchcock, *Architecture: Nineteenth and Twentieth Centuries*, 3rd ed. (Baltimore: Penguin Books, 1969), p. 385.

29. Council on Tall Buildings and Urban Habitat, *Architecture of Tall Buildings* (New York: McGraw-Hill, 1995), p. 4.

30. A. F. Burstall, *A History of Mechanical Engineering* (London: Faber and Faber, 1963).

31. Watt quoted in D. P. Miller, *History of Science* 38: 1 (2000).

32. Stephenson quoted in A. Pacey, *The Maze of Ingenuity* (Cambridge, Mass.: MIT Press, 1974). Compare the assessments of Burstall, *Mechanical Engineering*, p. 280, and T. S. Reynolds, *Technology and Culture* 20: 270 (1979).

33. For the water turbine, see N. A. F. Smith, *History of Technology* 2: 215 (1977). For internal combustion engines, see C. L. Cummins, Jr., *Internal Fire* (Lake Oswego, Ore.: Carnot Press, 1976); J. St. Peter, *The History of Aircraft Gas Turbine Engine Development in the United States* (Atlanta: International Gas Turbine Institute, 1999); and E. W. Constant II, *The Origins of the Turbojet Revolution* (Baltimore: Johns Hopkins University Press, 1980).

34. L. T. C. Rolt, *A Short History of Machine Tools* (Cambridge, Mass.: MIT Press, 1965). R. S. Woodbury, *Studies in the History of Machine Tools* (Cambridge, Mass.: MIT Press, 1972). J. F. Reintjes, *Numerical Control: Making a New Technology* (New York: Oxford University Press, 1991).

35. Fairbairn quoted in Rolt, *Machine Tools*, p. 91.

36. M. L. Dertouzos, R. K. Lester, and R. M. Solow, *Made in America* (New York: Harper, 1989), pp. 232, 233.

37. B. Douthwaite, *Enabling Innovation* (London: Zed Books, 2002), p. 37.

38. R. S. Woodbury, *Machine Tools*, p. 99.

39. O. Mayr, *The Origins of Feedback Control* (Cambridge, Mass.: MIT Press, 1970). S. Bennett, *A History of Control Engineering: 1800–1930* and *1930–1955* (London: IEE Press, 1979; 1993). D. A. Mindell, *Between Human and Machine* (Cambridge, Mass.: MIT Press, 2002). See also *IEEE Control Systems* 16 (3) (1996), a special historical issue.

40. Designation of the second industrial revolution is less clear than the first. I follow the definition of D. S. Landes, *The Unbound Prometheus* (New York: Cambridge University Press, 1969), p. 4.

41. D. C. Jackson, *Electrical Engineering* 53: 770 (1934).

42. W. H. Brock, *History of Chemistry* (New York: Norton, 1992). L. F. Haber, *The Chemical Industry during the Nineteenth Century* (New York: Oxford University Press, 1958), chap. 6. J. J. Beer, *Isis* 49, pt. 2 (156): 123 (1958). W. F. Furter, ed., *History of Chemical Engineering* (Washington D.C.: American Chemical Society, 1980). O. A. Hougen, *Chemical Engineering Progress* 73 (1): 89 (1977).

43. L. E. Scriven, *Advances in Chemical Engineering* 16: 3 (1991).

44. A. Arora, R. Landau, and N. Rosenberg, eds., *Chemicals and Long-term*

Economic Growth (New York: Wiley, 1998). R. Landau and N. Rosenberg, in *Technology and the Wealth of Nations,* ed. N. Rosenberg, R. Landau, and D. C. Mowery (Stanford: Stanford University Press, 1992), pp. 73–120. L. F. Haber, *The Chemical Industry: 1900–1930* (New York: Oxford University Press, 1971). K. Wintermantel, *Chemical Engineering Science* 54: 1601 (1999).

45. W. K. Lewis, *Chemical Engineering Progress Symposium Series* 55 (26): 1 (1959).

46. W. H. Walker, W. K. Lewis, and W. H. McAdams, *Principles of Chemical Engineering* (New York: McGraw-Hill, 1923), p. iii. McAdams quoted in D. A. Hounshell and J. K. Smith, *Science and Corporate Strategy* (New York: Cambridge University Press, 1988), p. 280.

47. Hougen, *Chemical Engineering Progress* 73 (1). Hounshell and Smith, *Corporate Strategy.* J. Ponton, *Chemical Engineering Science* 50: 4045 (1995).

48. R. L. Pigford, *Chemical and Engineering News,* Centennial Issue: 190–203 (April 1976).

49. A. L. Elder, *Chemical Engineering Progress Symposium* 66 (100) (1970). G. L. Hobby, *Penicillin: Meeting the Challenge* (New Haven: Yale University Press, 1985).

50. Arora, Landau, and Rosenberg, *Chemicals.* P. H. Spitz, *Petrochemicals: The Rise of an Industry* (New York: John Wiley and Sons, 1988). R. Landau, *Uncaging Animal Spirits: Essays in Engineering, Entrepreneurship, and Economics* (Cambridge, Mass.: MIT Press, 1994). Landau and Rosenberg, *Technology.* Pigford, *Chemical and Engineering News.* E. P. Kropp, *Chemical Engineering Progress* 93 (1): 42 (1997).

51. Quoted by E. J. Gornowski in Furter, *Chemical Engineering,* pp. 303–312.

52. Howard quoted in R. Landau, *Chemical Engineering Progress* 93 (1): 52 (1997).

53. P. Dunsheath, *A History of Electrical Power Engineering* (Cambridge, Mass.: MIT Press, 1962). R. Rosenberg, *IEEE Spectrum* 21 (7): 60 (1984). K. L. Wildes and N. A. Lindgren, *A Century of Electrical Engineering and Computer Science at MIT: 1882–1982* (Cambridge, Mass.: MIT Press, 1985). J. D. Ryder and D. G. Fink, *Engineers and Electrons* (New York: IEEE Press, 1984).

54. T. P. Hughes, *Networks of Power* (Baltimore: Johns Hopkins University Press, 1985).

55. J. Douglas, *EPRI Journal* 24 (2): 18 (1999). T. J. Overbye and J. D. Weber, *IEEE Spectrum* 38 (2): 52 (2001). J. Makansi, *IEEE Spectrum* 38 (2): 24 (2001). H. B. Püttgen, D. R. Volzka, and M. I. Olken, *IEEE Power Engineering Review* 21 (2): 8 (2001).

56. A. Lurkis, *The Power Brink* (New York: Icare Press, 1982), p. 51.

57. Aeschylus, *Agamemnon,* 282. This 458 B.C.E. drama was written seven or eight centuries after the Trojan War.

58. S. Leinwoll, *From Spark to Satellite* (New York: Charles Scribner's Sons, 1979). J. Bray, *The Communications Miracle* (New York: Plenum, 1995). L. Solymar, *Getting the Message* (New York: Oxford University Press, 1999). Many historical papers are found in the Institute of Electrical Engineers, *100 Years of Radio,* IEE Conference Publication 411 (1995); the July 1998 issue of *Proceedings of the IEEE,* commemorating the mobile radio centennial; and the May 2002 issue of *IEEE Communications Magazine,* celebrating its fiftieth anniversary.

3. ENGINEERING FOR INFORMATION

1. F. Seitz, *Electronic Genie* (Urbana: University of Illinois Press, 1998), p. 1.

2. W. Shockley, *IEEE Transactions on Electron Devices* ED-23: 597 (1976). M. Riordan and L. Hoddeson, *Crystal Fire* (New York: Norton, 1997). C. M. Melliar-Smith, *Proceedings of the IEEE* 86: 86 (1998). W. F. Brinkman and D. V. Lang, *Review of Modern Physics* 71: S480 (1999).

3. The positively charged carrier of current, called a "hole," is actually a vacancy in a sea of electrons.

4. Shockley, *IEEE Transactions* ED-23.

5. J. S. Kilby, *IEEE Transactions on Electron Devices* ED-23: 648 (1976). I. M. Ross, *Proceedings of the IEEE* 86: 7 (1998). C. T. Sah, ibid. 76: 1280 (1988). G. R. Moore, ibid., 86: 53 (1998). D. G. Rea et al., *Research Technology Management* 40 (4): 46 (1997).

6. G. E. Moore in *Engines of Innovation,* ed. R. S. Rosenbloom and U. J. Spencer, eds. (Cambridge, Mass.: Harvard Business School Press, 1996), pp. 135–147.

7. R. K. Lester, *The Productive Edge* (New York: Norton, 1998), chap. 3.

8. Kilby, *IEEE Transactions* ED-23.

9. G. E. Moore, *Electronics,* April 19, 1965, pp. 114–117.

10. Moore, *Electronics;* G. E. Moore, *Optical/Laser Microlithography VIII: Proceedings of SPIE* 2440: 2 (1995).

11. B. T. Murphy, D. E. Haggan, and W. W. Troutman, *Proceedings of the IEEE* 88: 691 (2000).

12. P. Gargini, *SPIE 2002 Microlithography Symposium* (2002).

13. See the special issue on semiconductor technology in *Proceedings of the IEEE* 89(3), 2001. See also the 2000 International Technology Roadmap of Semiconductors, prepared by industry associations in America, Asia, and Europe, and available at www.public.itrs.net.

14. R. P. Feynman, in *Miniaturization,* ed. H. D. Gilbert (New York: Reinhold, 1959).

15. M. Gross, *Travels into the Nanoworld* (Cambridge: Perseus, 1999). Comprehensive reviews and visions of science and technology are available at itir.loyola.edu/nano/TWGN.Research.Direction and www.nano. gov. R. Compano, *Nanotechnology* 12: 85 (2001) is a concise review.

16. S. T. Picraux and P. J. McWhorter, *IEEE Spectrum* 35 (12): 24 (1998). Craighead, *Science,* 290, 1532 (2000). F. B. Prinz, A. Golnas, and A. Nickel, *MRS Bulletin,* 25, 32 (2000). T. Chovan and A. Guttman, *Trends in Biotechnology,* 20(3), 116 (2002).

17. Y. Wada, *Proceedings of the IEEE* 89: 1147 (2001). G. M. Whitesides and B. Grzybowski, *Science* 295: 2418 (2002).

18. M. S. Dresselhaus, G. Dresselhaus, and P. Avouris, eds., *Carbon Nanotubes: Synthesis, Structure, Properties, and Applications* (Berlin: Springer, 2001). R. H. Baughman, A. A. Zakhidov, and W. A. de Heer, *Science* 297: 787 (2002).

19. J. Chabert, *A History of Algorithms* (Berlin: Springer, 1999). C. B. Boyer, *A History of Mathematics* (Princeton: Princeton University Press, 1985), chap. 13.

20. M. Campbell-Kelly and W. Aspray, *Computer: A History of the Information Machine* (New York: Basic Books, 1996). P. E. Ceruzzi, *A History of Modern Computing* (Cambridge, Mass.: MIT Press, 1998). G. Ifrah, *The Universal History of Computing* (New York: Wiley, 2001).

21. J. van der Spiegel et al. in *The First Computers,* ed. R. Rojas and O. Hashagen (Cambridge, Mass.: MIT Press, 2000), pp. 121–189.

22. S. McCartney, *ENIAC* (New York: Walker, 1999).

23. M. V. Wilkes, *Memories of a Computer Pioneer* (Cambridge, Mass.: MIT Press, 1985), p. 123.

24. J. E. Hopcroft, *Annual Review of Computer Science* 4: 1 (1990).

25. S. V. Pollack, in *Studies in Computer Science,* ed. S. V. Pollack (Washington D.C.: Mathematical Association of America, 1982), pp. 1–51. R. W. Hamming, in *ACM Turing Award Lectures* (New York: ACM Press, 1968), pp. 207–218. F. Brooks, *Communications of the ACM* 39 (3): 61 (1996).

26. P. J. Dennings et al., *Communications of the ACM* 32 (1): 9 (1989). J. Hartmanis, *ACM Computing Surveys* 27: 7 (1995). M. C. Loui, ibid., 31.

27. The top ten are Carnegie-Mellon; MIT; Stanford; University of California, Berkeley; University of Illinois, Urbana-Champaign; University of Texas, Austin; University of Washington; University of Michigan, Ann Arbor; Princeton; and Cornell. Source: www.usnews.com.

28. ACM's 1998 computing classification system includes eleven large areas: general literature, hardware, computer systems organization, software, data, theory of computation, mathematics of computing, information systems, computing methodologies, computing applications, computing milieu. ACM and IEEE-CS's computing curriculum for 2001 lists four-

teen knowledge focus groups: discrete structures, programming fundamentals, algorithms and complexity, programming languages, architecture and organization, operating systems, net-centric computing, human-
computer interface, graphics and visual computing, intelligent systems,
information management, software engineering, social and professional
issues, computational science.

29. J. Hartmanis and H. Lin, eds., *Computing the Future* (Washington D.C.:
National Academy Press, 1992).

30. A. Sameh, *ACM Computing Surveys* 28: 810 (1996).

31. R. E. Smith, *IEEE Annals of the History of Computing* 10: 277 (1989).

32. A. D. Tanenbaum, *Structured Computer Organization,* 4th ed. (Upper
Saddle River, N.J.: Prentice Hall, 1999), p. 8.

33. N. Tredennick, *IEEE Computer* 29 (10): 27 (1996).

34. G. M. Hopper and J. W. Mauchly, *Proceedings of the IRE* 40: 1250
(1953). Compare the philosophy of H. Putnam, *Mind, Language, and
Reality* (New York: Cambridge University Press, 1975), chap. 18, and the
engineering wisdom in F. Brooks, *The Mythical Man-Month,* 2nd ed.
(New York: Addison-Wesley, 1995).

35. P. Wegner, *IEEE Transaction on Computers* 25: 1207 (1976).

36. D. E. Knuth and L. T. Pardo, in *A History of Computing in the Twentieth Century,* ed. N. Metropolis, J. Howlett, and G. Rota (New York:
Academic Press, 1980), pp. 197–274. H. Backus's article is on pp. 125–
136.

37. S. Rosen, *Communications of the ACM* 15 (7): 591 (1972).

38. D. M. Ritchie, in *Great Papers in Computer Science,* ed. P. Laplante
(New York: IEEE Press, 1996), pp. 705–717.

39. R. Comerford, *IEEE Spectrum* 36 (5): 25 (1999). J. Kerstetter,
BusinessWeek, March 3, 2003.

40. J. Hennessy, *IEEE Computer* 32 (8): 27 (1999).

41. C. E. Shannon, *Bell System Technical Journal* 27: 379–423, 623–656
(1948).

42. J. Bray, *The Communications Miracle* (New York: Plenum, 1995).
L. Solymar, *Getting the Message* (New York: Oxford University Press,
1999).

43. J. Coopersmith, *IEEE Spectrum* 30 (2): 46 (1993).

44. J. R. Pierce, *The Beginning of Satellite Communication* (San Francisco:
San Francisco Press, 1968). P. T. Thompson and D. Grey, in *100 Years of
Radio,* IEE Conference Publication 411 (1995), pp. 199–206. W. Wu,
Proceedings of the IEEE 85: 998 (1997). A. Jamalipour and T. Tung,
IEEE Personal Communications 8 (3): 5 (2001).

45. N. Holonyak, *Proceedings of the IEEE* 85: 1678 (1997).

46. J. Hecht, *City of Light* (New York: Oxford University Press, 1999). C. D.
Chaffee, *The Rewiring of America* (Boston: Academic Press, 1988).

47. N. S. Bergano, *Optics and Photonics News* 11 (3): 20 (2000).

48. A. Dutta-Roy, *IEEE Spectrum* 36 (3): 32 (1999). E. Hurley and J. H. Keller, *The First 100 Feet* (Cambridge, Mass.: MIT Press, 1999).

49. T. S. Rappaport, A. R. M. Buehrer, and W. H. Tranter, *IEEE Communications Magazine* 40 (5): 148 (2002). D. C. Cox, *IEEE Personal Communications* (4): 20 (1995). M. W. Oliphant, *IEEE Spectrum* 36 (8): 20 (1999).

50. J. Adam, *IEEE Spectrum* 33 (9): 57 (1996).

51. B. M. Leiner et al., *Communications of the ACM* 40 (2): 102 (1997). K. Hafner and M. Lyon, *Where Wizards Stay Up Late* (New York: Touchstone, 1996). J. Abbate, *Inventing the Internet* (Cambridge, Mass.: MIT Press, 1999). See also "Internet: Past, Present, Future," in the July 2002 issue of *IEEE Communications Magazine*.

52. B. White, *Physics Today* 51 (11): 30 (1998). J. Gillies and R. Cailliau, *How the Web Was Born* (New York: Oxford University Press, 2000).

4. ENGINEERS IN SOCIETY

1. L. Marx, *Social Research* 64: 965 (1997).

2. J. P. Richter, ed. *The Notebooks of Leonardo da Vinci* (New York: Dover, 1970), pp. 14–15, 328.

3. H. Hoover, *The Memoirs of Herbert Hoover* (New York: Macmillan, 1951).

4. Roosevelt's letter, dated January 1906, is quoted by D. McCullough in *Sons of Martha: Civil Engineering Readings in Modern Literature,* ed. A. J. Fredrich (New York: ASCE Press, 1978), pp. 587–594.

5. R. P. Multhauf, *Technology and Culture* 1: 38 (1959).

6. A. E. Musson and E. Robinson, *Science and Technology in the Industrial Revolution* (Toronto: University of Toronto Press, 1969), pp. 72–73.

7. Ibid., pp. 73–75.

8. S. Pollard, *Britain's Prime and Britain's Decline* (London: Edward Arnold, 1989), p. 127.

9. W. H. G. Armytage, *A Social History of Engineering* (London: Faber and Faber, 1976). J. B. Rae and R. Volti, *The Engineer in History* (New York: Peter Lang, 1993). P. Elliott, *Annals of Science* 57: 61 (2000).

10. The text of the Royal Charter was written by Thomas Tredgold. For engineering societies, see P. Lundgreen, *Annals of Science* 47: 33 (1990). T. S. Reynolds, ed., *The Engineer in America* (Chicago: University of Chicago Press, 1991).

11. Quoted in D. S. L. Cardwell and R. L. Hills, *History of Technology* 1: 1 (1976).

12. J. H. Weiss, *The Making of Technological Man* (Cambridge, Mass.: MIT Press, 1982). D. O. Belanger, *Enabling American Innovation* (West Lafayette, Ind.: Purdue University Press, 1998).

13. Both remarks quoted in R. L. Geiger, *To Advance Knowledge* (New York: Oxford University Press, 1986), pp. 13–14.

14. C. Tichi, *Shifting Gear* (Chapel Hill: University of North Carolina Press, 1987), pp. 98–99, 119–120.

15. B. Sinclair, in *American Technology,* ed. C. Pursell (Malden, Mass.: Blackwell, 2001), pp. 145–154. Tichi, *Shifting Gear.*

16. Marx, *Social Research* 64.

17. Quoted by G. L. Downey and J. C. Lucena, in *Handbook of Science and Technology Studies,* ed. S. Jasanoff et al. (Thousand Oaks, Calif.: Sage, 1995), pp. 167–188.

18. Sinclair, *American Technology.*

19. Quoted by R. C. Maclaurin, in *Technology and Industrial Efficiency,* MIT ed. (New York: McGraw-Hill, 1911), pp. 1–10.

20. L. E. Grinter, *Journal of Engineering Education* 44: 25 (1955).

21. The 1997–98 mean GRE verbal scores for U.S. citizens in various fields were: engineering, 499; computer science, 516; biological sciences, 507; physical sciences, 510; behavioral sciences, 488; social sciences, 473; all fields, 481. http://208.249.124.108/web/site/bbcharts/bbs.htm.

22. R. P. Feynman, *What Do You Care What Other People Think?* (New York: Bantam, 1988), p. 184; *Surely You're Joking, Mr. Feynman!* (New York: Bantam, 1985), p. 256.

23. E. Ashby, *Technology and the Academics* (London: Macmillan, 1959), p. 66.

24. C. P. Snow, *The Two Cultures and a Second Look* (New York: Cambridge University Press, 1963).

25. Quoted in W. Symonds, *BusinessWeek,* February 18, 2002, pp. 72–78.

26. P. Gross, N. Levitt, and M. Lewis, eds. *The Flight from Science and Reason* (Baltimore: Johns Hopkins, 1996). A. Ross, ed., *Science Wars* (Durham: Duke University Press, 1996).

27. T. S. Reynolds, *Technology and Culture* 42: 523 (2001).

28. H. A. Bauer, in *Beyond the Science Wars,* ed. U. Segerstråle (Albany: State University of New York Press, 2000), pp. 41–62.

29. U. Segerstråle, in Segerstråle, *Science Wars,* p. 6.

30. A. Roland, *Technology and Culture* 38: 697 (1997).

31. John Rae quoted in Reynolds, *Technology and Culture* 42.

32. A 2002 ITEA/Gallup poll (www.iteawww.org) found that only 1 percent of respondents thought it is not very important for people at all levels to develop some ability to understand and use technology, while 75 percent said they would like to know how technologies work.

33. M. R. Smith and G. Clancey, eds., *Major Problems in the History of American Technology* (Boston: Houghton Mifflin, 1998). The editors wrote: "We have purposely limited our discussion of the history of engineering, because the story has been so well told elsewhere," without hint-

ing as to where or explaining why equally well-represented fashionable topics were included.

34. J. M. Staudenamier, *Technology's Storytellers* (Cambridge, Mass.: MIT Press, 1985).

35. T. P. Hughes, *Technology and Culture* 22: 550 (1981).

36. J. R. Cole, *The Bridge* 26 (3): 1 (1996). Such criticisms have effects. The Sloan Foundation commissioned a text on American history that includes technology, *Inventing America*.

37. C. M. Vest, in *AAAS Science and Technology Policy Yearbook* 2000, www.aaas.org/spp/yearbook/2000/ch28.

38. The identifying of technology with computer is repeated three times and the disconnect between technology and engineering twice in R. Williams, *Technology and Culture* 41: 641 (2000), which is quoted. They are repeated in Williams's book *Retooling* (Cambridge, Mass.: MIT Press, 2002). Is engineering really disconnected from technology, as alleged, or does the allegation reveal the disconnect of both from technology *studies?* Decide for yourself by checking the lecture and conference calendars on MIT's website. Read the descriptions of talks that have "technology" in their title—there are usually some there—and see if they are all computer and no engineering. When I looked at them for the week of February 12, 2001, I found three talks: "An Assessment of Future Automotive Technology," "Missile Defense, Technology and Policy," "How Technology Empowers Imagination."

39. M. R. C. Greenwood and K. K. North, *Science* 286: 2072 (1999).

40. M. Appl, in *A Century of Chemical Engineering*, ed. W. F. Furter (Plenum, New York, 1982), pp. 29–54.

41. B. Seely, *Technology and Culture* 34: 344 (1993). D. F. Noble, *America by Design* (Oxford University Press, New York, 1977). Belanger, *American Innovation*. R. Locke, in *Managing in Different Cultures*, P. Joynt and M. Warner, ed. (Universitetsforlaget, Oslo, Norway, 1985), pp. 166–216.

42. J. W. Servos, *Isis* 71: 531 (1980). Geiger, *To Advance Knowledge*, pp. 179–183.

43. W. Wickenden, *Mechanical Engineering* 51: 586 (1929).

44. G. W. Matkin, *Technology Transfer and the University* (New York: Macmillan International, 1990).

45. A. M. McMahon, *The Making of a Profession* (New York: IEEE Press, 1984), pp. 68, 76–78. Belanger, *American Innovation*, pp. 13–15. Geiger, *To Advance Knowledge*, p. 181.

46. J. A. Armstrong, in *Forces Shaping the U.S. Academic Engineering Research Enterprise*, National Academy of Engineering (Washington, D.C.: National Academy Press, 1995), pp. 59–68.

47. F. E. Terman, *Proceedings of the IRE* 50: 955 (1962).

48. F. E. Terman, *Proceedings of the IEEE* 64: 1399 (1976). W. R. Perkins, ibid. 86: 1788 (1998).

49. National Science Foundation, *Science and Engineering Indicators 2002,* pp. 4–10.

50. W. R. Whitney, in *Technology and Industrial Efficiency,* pp. 80–89.

51. L. S. Reich, *The Making of American Industrial Research* (New York: Cambridge University Press, 1985). M. Crow and B. Barry, *Limited by Design* (New York: Columbia University Press, 1998). R. Buderi, *Engines of Tomorrow* (New York: Simon and Schuster, 2000). L. Geppert, *IEEE Spectrum* 31 (9): 30 (1994).

52. H. Ernst, C. Leptien, and J. Vitt, *IEEE Transactions on Engineering Management* 47: 184 (2000).

53. Quoted in Reich, *Industrial Research,* p. 37.

54. Armstrong quoted in Buderi, *Engines of Tomorrow,* p. 129. For Xerox's case, see M. B. Myers and R. S. Rosenbloom, *Research Technology Management* 39 (3): 14 (1996).

55. R. E. Gomory, *Research Technology Management* 32 (6): 27 (1989). L. S. Edelheit, ibid. 41 (2): 21 (1998). Buderi, *Engines of Tomorrow.*

56. Census Bureau, *Statistical Abstract of the United States 2001,* Table 1269.

57. In R. A. Dawe, ed., *Modern Petroleum Technology* (New York: Wiley, 2000), p. xiv.

58. *Time's* online poll on "the event of the century" was an unscientific informal survey that ended on January 19, 2000. Its results are posted at www.time.com/time/time100/t100events.htm. The top twenty nominees were: (1) Elvis teaches American teens to rock and roll, (2) first landing on the moon, (3) Gandhi opposes Britain with civil disobedience, (4) World War II, (5) U.S. civil rights movement, (6) the Holocaust, (7) invention of the microchip, (8) Internet created, (9) Model T Ford introduced, (10) theory of relativity presented, (11) first atomic bomb dropped, (12) World War I, (13) first electronic computer unveiled, (14) first radio signal broadcast, (15) Berlin Wall falls, (16) invention of the airplane, (17) invention of the transistor, (18) Russian Revolution, (19) first nuclear chain reaction, (20) Soviet Union dissolves.

59. N. Armstrong, *The Bridge* 30 (1): 15 (2000), is also available at www.greatachievements.org. The National Academy of Engineering invited twenty-nine discipline-specific engineering societies to submit nominations; the winners were chosen from 105 nominees by a committee of academy members. The 20 selected as the greatest engineering achievements of the century were: (1) electrification, (2) the automobile, (3) the airplane, (4) safe and abundant water, (5) electronics, (6) radio and television, (7) agricultural mechanization, (8) computers, (9) the telephone, (10) air conditioning and refrigeration, (11) interstate highways, (12)

space exploration, (13) the Internet, (14) imaging technologies, (15) household appliances, (16) health technologies, (17) petroleum and gas technologies, (18) laser and fiber optics, (19) nuclear technologies, (20) high-performance materials.

60. V. Smil, *Annual Review of Energy and Environment* 25: 21 (2000).

61. R. N. Anderson, *Scientific American* 278 (3): 86 (1998). Dawe, *Petroleum Technology.*

62. M. I. Hoffert et al., *Science* 298: 981–988 (2002).

63. S. R. Bull, *Proceedings of the IEEE* 89: 1216–1227 (2001).

64. R. Mandelbaum, *IEEE Spectrum* 39 (10): 34 (2002).

65. R. F. Service, *Science* 288: 1955 (2000). M. A. Weiss et al. *On the Road in 2020,* Energy Laboratory Report # MIT EL-00–003, web.mit.edu/energylab/www/.

66. N. Rosenberg, *Inside the Black Box* (Cambridge University Press, New York, 1982), Ch.3.

67. D. N. Ghista, *IEEE Engineering in Medicine and Biology* 19 (6): 23 (2000). F. Nebekee, ibid. 21 (3): 17 (2002).

68. A. Lawler, *Science* 288: 32 (2000).

5. INNOVATION BY DESIGN

1. M. J. Seifer, *Wizard* (Secaucus N.J.: Birch Lane Press, 1996), p. 23.

2. T. D. Crouch, *The Bishop's Boys* (New York: Norton, 1989), p. 228.

3. J. Wiesner, in *Listen to Leaders in Engineering,* ed. A. Love and J. S. Childers (Atlanta: Tupper and Love, 1965), pp. 323–338.

4. G. Stix, *IEEE Spectrum* 25: 76 (1988).

5. H. Ford, *My Life and Work* (New York: Doubleday, Page and Co., 1922), p. 30.

6. S. C. Florman, *The Existential Pleasure of Engineering* (New York: St. Martin's, 1976).

7. C. E. Shannon, in *Claude Elwood Shannon: Miscellaneous Writings,* ed. N. J. A. Sloane and A. D. Wyner (New York: IEEE Press, 1993), #72.

8. R. P. Feynman, *What Do You Care What Other People Think?* (New York: Bantam, 1988), p. 243.

9. J. P. Richter, ed., *The Notebooks of Leonardo da Vinci* (New York: Dover, 1970), p. 18.

10. W. Wordsworth, 1802 preface in *Lyrical Ballads,* ed. W. J. B. Owen (New York: Oxford University Press, 1969), p. 157.

11. J. W. von Goethe, "Nature and Art," in *German Poetry from 1750–1900,* ed. R. M. Browning (New York: Continuum, 1984), p. 59.

12. Einstein's 1952 letter to M. Solovine appears in *Einstein,* ed. A. P. French (Cambridge, Mass.: Harvard University Press, 1979), pp. 269–271.

13. D. Slepian, *Proceedings of the IEEE* 64: 272 (1976).

14. Feynman, *What Do You Care*, p. 245.

15. A. Einstein, *Ideas and Opinions* (New York: Crown, 1954), p. 343.

16. Einstein, *Ideas*, p. 266.

17. G. Pólya, *Mathematics and Plausible Reasoning* (Princeton: Princeton University Press, 1954), p. vi.

18. H. Petroski, *Invention by Design* (Harvard University Press, Cambridge, 1996), p. 2.

19. G. Galilei, *Dialogue Concerning the Two Chief World Systems* (University of California Press, Berkeley, 1967), p. 341.

20. M. Polanyi, *Personal Knowledge* (Chicago: University of Chicago Press, 1958), p. vii.

21. Pólya, *Mathematics*.

22. M. W. Maier and E. Rechtin, *The Art of Systems Architecting* (Boca Raton, Fla.: CRC Press, 2000), pp. 28–29.

23. S. Y. Auyang, *Foundations of Complex-System Theories* (New York: Cambridge University Press, 1998), chap. 3.

24. Quoted in M. Josephson, *Edison* (New York: McGraw-Hill, 1959), p. 198.

25. R. P. Feynman, *The Character of Physical Law* (Cambridge, Mass.: MIT Press, 1965), p. 164.

26. W. Heisenberg, *Tradition in Science* (New York: Seabury Press, 1983), p. 128.

27. H. S. Black, *IEEE Spectrum* 14 (12): 55 (1977). H. W. Bode, in *Selected Papers on Mathematical Trends in Control Theory*, ed. R. Bellman and R. Kalaba (New York: Dover, 1960), pp. 106–123. S. Bennett, *A History of Control Engineering: 1930–1955* (London: IEE Press, 1993), chap. 3.

28. Black, *IEEE Spectrum* 14 (2).

29. Shannon, *Miscellaneous Writings*, #72.

30. "Rules of reasoning" from Newton's *Principia*, Book III, repr. in *Newton's Philosophy of Nature*, ed. H. S. Thayer (New York: Hafner, 1953), p. 3.

31. Einstein, *Ideas*, p. 272.

32. Ford, *My Life*, pp. 13–14.

33. C. Murray and C. B. Cox, *Apollo* (New York: Simon and Schuster, 1989), pp. 175–176.

34. Shannon, *Miscellaneous Writings*, #72.

35. D. P. Billington, *The Tower and the Bridge* (Princeton: Princeton University Press, 1983).

36. Seifer, *Wizard*, p. 25.

37. Plato, *Republic*, 368–369.

38. Stix, *IEEE Spectrum* 25.

39. R. F. Miles, ed. *Systems Concepts* (Wiley, New York, 1973), p. 11.

40. Einstein, *Ideas*, p. 324.

41. See *system* in the *Oxford English Dictionary.*

42. J. F. McCloskey, *Operations Research* 35: 143, 910 (1987). A. C. Hughes and T. P. Hughes, eds. *Systems, Experts, and Computers* (Cambridge, Mass.: MIT Press, 2000).

43. Bennett, *Control Engineering,* p. 164.

44. M. D. Fagen, ed., *National Service in War and Peace* (Murray Hill, N.J.: Bell Telephone Laboratory, 1979), pp. 618–619.

45. Bennett, *Control Engineering,* p. 204. For the history of systems engineering, see J. H. Brill, *Systems Engineering* 1: 258 (1998), and M. Kayton, *IEEE Transactions on Aerospace and Electronic Systems* 33: 579 (1997).

46. T. P. Hughes, *Rescuing Prometheus* (New York: Pantheon Books, 1998).

47. S. Ramo, in Miles, *Systems Concepts,* pp. 13–32.

48. R. P. Smith, *IEEE Transactions on Engineering Management* 44: 67 (1997).

49. Ramo, in Miles, *Systems Concepts.*

50. A. Rosenblatt and G. F. Watson, *IEEE Spectrum* 28 (7): 22 (1991).

51. S. Shapiro, *IEEE Annals of the History of Computing* 19: 20 (1997). P. Bourque et al., *IEEE Software* 16 (6): 35 (1999). R. H. Thayer, *Computer* 35 (4): 68 (2002).

52. Quoted in R. R. Schaller, *IEEE Spectrum* 34 (6): 53 (1997).

53. J. Horning, *Communications of the ACM* 44 (7): 112 (2001).

54. M. Keil et al., *Communications of the ACM* 41 (11): 76 (1998).

55. A. Rosenblatt and G. F. Watson, *IEEE Spectrum* 28 (7): 22 (1991).

56. D. C. Aronstein and A. C. Piccirillo, *Have Blue and the F-117A* (Reston, Va.: AIAA, 1997). K. Sabbagh, *21st-Century Jet* (London: Macmillan, 1995). G. Norris, *IEEE Spectrum* 32 (10): 20 (1995).

57. B. R. Rich and L. Janos, *Skunk Works* (Boston: Little, Brown, 1994), p. 115.

58. K. Forsberg and H. Mooz, in *Software Requirements Engineering,* 2nd ed., ed. R. H. Thayer, M. Dorfman, and A. M. Davis (Los Alamitos, Calif.: IEEE Computer Society Press, 1997), pp. 44–72.

59. R. G. O'Lone, *Aviation Week & Space Technology* 134 (22): 34 (June 3, 1991). P. Proctor, ibid., 140 (5): 37 (April 11, 1994).

60. P. E. Gartz, *IEEE Transactions on Aerospace and Electronic Systems* 33: 632 (1997).

61. P. M. Condit, *Research Technology Management* 37 (1): 33 (1994).

62. Ibid. Petroski, *Invention by Design,* chap. 7.

63. A. L. Battershell, *The DoD C-17 versus the Boeing 777* (Washington, D.C.: National Defense University, 1999).

64. J. Lovell and J. Kluger, *Apollo 13* (Boston: Houghton Mifflin, 1994).

65. Sabbagh, *21st-Century Jet,* p. 75.

66. Forsberg and Mooz, *Software Requirements.*

67. H. Buus et al., *IEEE Transactions on Aerospace and Electronic Systems* 33: 656 (1997).

68. Aronstein and Piccirillo, *Have Blue,* p. 194.

69. F. Brooks, *The Mythical Man-Month,* 2nd ed. (New York: Addison-Wesley: 1995), p. 184.

70. R. Bell and P. A. Bennett, *Computing and Control Engineering Journal* 11 (1): 3 (2000).

71. G. Stix, *Scientific American* 271 (5): 96 (1994). U.S. General Accounting Office, *Evolution and Status of FAA's Automation Program,* GAO/T-RCED/AIMD-98-85 (1998).

72. Aronstein and Piccirillo, *Have Blue,* p. 157; see also pp. 61–62.

73. J. E. Steiner, *Case Study in Aircraft Design* (Reston, Va.: AIAA, 1978), p. 71.

74. Condit, *Research Technology Management* 37 (1).

75. W. G. Vincenti, *What Engineers Know and How They Know It* (Baltimore: Johns Hopkins University Press, 1990), chap. 3. L. Adelman, *IEEE Transactions on Systems, Men, and Cybernetics* 19: 483 (1989).

76. Rich and Janos, *Skunk Works,* p. 88.

77. O'Lone, *Aviation Week* 134 (22). Petroski, *Invention by Design.*

78. Gartz, *IEEE Transactions on Aerospace and Electronic Systems* 33.

79. Aronstein and Piccirillo, *Have Blue,* pp. 36, 175. Rich and Janos, *Skunk Works,* pp. 47–48.

80. Aronstein and Piccirillo, *Have Blue,* pp. 161–162.

81. Rich and Janos, *Skunk Works,* pp. 88, 332–333.

82. Interview in Sabbagh, *21st-Century Jet,* pp. 63–64.

83. Steiner, *Case Study,* p. 7. Condit, *Research Technology Management* 37 (1).

84. Interview in Sabbagh, *21st-Century Jet,* p. 64.

85. Brooks, *Mythical Man-Month,* p. 143.

86. Auyang, *Complex-System Theories,* section 6.

87. Plato, *Phaedrus,* 265–266.

88. W. B. Parsons, *Engineers and Engineering in the Renaissance* (Cambridge, Mass.: MIT Press, 1939), p. 25.

89. W. A. Wallace, *Galileo and His Sources* (Princeton: Princeton University Press, 1984), p. 119.

90. Query 31 of Newton's *Opticks,* reprinted in Thayer, *Newton's Philosophy,* pp. 178–179.

91. J. Cottingham, R. Stoothoff, and D. Murdoch, eds., *The Philosophical Writings of Descartes* (New York: Cambridge University Press, 1985), vol. 1, p. 20.

92. Vincenti, *What Engineers Know,* p. 9.

93. Gartz, *IEEE Transactions on Aerospace and Electronic Systems* 33. Sabbagh, *21st-Century Jet*, pp. 72–73.

94. Forsberg and Mooz, *Software Requirements*.

95. B. Witwer, *IEEE Transactions on Aerospace and Electronic Systems* 33: 637 (1997). S. L. Pelton and K. D. Scarbrough, ibid. 33: 642.

96. Proctor, *Aviation Week* 134 (22). Norris, *IEEE Spectrum* 32 (10). Witwer, *IEEE Transactions on Aerospace and Electronic Systems* 33.

97. Sabbagh, *21st-Century Jet,* pp. 89–90.

98. W. Vincenti, *Technology and Culture* 35: 1 (1994); *Social Studies of Science* 25: 553 (1995).

99. Rich and Janos, *Skunk Works,* p. 225. Pelton and Scarbrough, *IEEE Transactions on Aerospace and Electronic Systems* 33.

100. Aronstein and Piccirillo, *Have Blue,* p. 232.

101. Brooks, *Mythical Man-Month.*

6.   SCIENCES OF USEFUL SYSTEMS

1. L. E. Grinter, *Journal of Engineering Education* 44: 25 (1955).

2. J. P. Richter, ed., *The Notebooks of Leonardo da Vinci* (New York: Dover, 1970), p. 11.

3. S. Drake, ed., *Discoveries and Opinions of Galileo* (New York: Doubleday, 1957), p. 238.

4. E. Wigner, *Symmetries and Reflections* (Cambridge, Mass.: MIT Press, 1967), p. 222.

5. R. Descartes, *The Philosophical Writings of Descartes,* ed. J. Cottingham, R. Stoothoff, and D. Murdoch (New York: Cambridge University Press, 1985), vol. 1, p. 19. Rules 4 and 16 of his *Rules for the Direction of the Mind* address the nature of mathematics.

6. Quoted in H. T. Davis, *The Theory of Linear Operators* (Bloomington, Ind.: Principia Press, 1936), p. 10.

7. *Computing in Science and Engineering* 2 (1): 22–79 (2000).

8. D. N. Rockmore, *Computing in Science and Engineering* 2 (1): 60 (2000).

9. Davis, *Linear Operators,* p. 7.

10. In Maxwell's *A Treatise on Electricity and Magnetism,* the chapter "General Equations of the Electromagnetic Field" lists thirteen sets of equations in terms of potentials. Eight of the thirteen Heaviside combined into four in terms of electric and magnetic fields, and we now call them Maxwell's equations.

11. N. Wiener, *Invention* (Cambridge, Mass.: MIT Press, 1954), pp. 69–76. P. J. Nahn, *Oliver Heaviside* (Baltimore: Johns Hopkins University Press, 1988).

12. For an appreciation of Heaviside, applied mathematician Edmund

Whittaker wrote: "Looking back on the controversy after thirty years, we should now place Operational Calculus with Poincaré's discovery of automorphic functions and Ricci's discovery of the Tensor Calculus as the three most important mathematical advances of the last quarter of the nineteenth century." *Bulletin of the Calcutta Mathematical Society* 20: 199–220 (1928). See also the Heaviside biography by Ernst Weber in O. Heaviside, *Electromagnetic Theory* (New York: Dover, 1893), pp. xv–xvii.

13. N. Bourbaki, *Elements of the History of Mathematics* (Berlin: Springer-Verlag, 1984), p. 21.

14. F. Klein, *Development of Mathematics in the 19th Century* (Brookline, Mass.: Math SCI Press, 1928), p. 48.

15. Heaviside, *Electromagnetic Theory*, sections 224, 437.

16. R. P. Feynman, *Surely You're Joking, Mr. Feynman!* (New York: Bantam, 1985), p. 225.

17. J. E. Bailey, *Biotechnology Progress* 14: 8 (1998).

18. R. E. Kalman, P. L. Falb, and M. A. Arbib, *Topics in Mathematical System Theory* (New York: McGraw-Hill, 1969), p. 27.

19. R. G. Gallager, *IEEE Transactions on Information Theory* 47: 2681 (2001).

20. T. Kailath, in *Communications, Computation, Control, and Signal Processing,* ed. A. Paulraj, W. Raychowdbury, and C. D. Schaper (Boston: Kluwer, 1997), pp. 35–65.

21. D. S. Bernstein, *IEEE Control Systems* 18 (2): 81 (1998).

22. L. A. Zadeh, *Proceedings of the IRE* 50: 856 (1962).

23. I. A. Getting, in *Theory of Servomechanisms,* ed. H. M. James, N. B. Nichols, and R. S. Phillip (New York, McGraw-Hill, 1947), pp. 1–22.

24. H. J. Sussmann and J. C. Willems, *IEEE Control Systems* 17 (3): 32 (1997).

25. Gallager, *IEEE Transactions on Information Theory* 47. See also J. R. Pierce, ibid. 19: 3 (1973). S. Verdú, ibid. 44: 2057 (1998). W. Gappmair, *IEEE Communications Magazine* 37 (4): 102 (1999).

26. C. E. Shannon, *Bell System Technical Journal* 27: 379, 623 (1948).

27. D. J. Costello et al., *IEEE Transactions on Information Theory* 44: 2531 (1998). E. Biglieri and P. D. Torino, *IEEE Communications Magazine* 49 (5): 128 (2002).

28. Pierce, *IEEE Transactions on Information Theory* 19.

29. D. Drajic and D. Bajic, *IEEE Communications Magazine* 49 (6): 124 (2002).

30. Verdú, *IEEE Transactions on Information Theory* 44. Gallager, ibid. 47.

31. Kailath, *Communications.* H. W. Sorenson, *IEEE Spectrum* 7 (7): 63 (1970).

32. J. R. Cloutier, J. H. Evers, and J. J. Feeley, *IEEE Control System Magazine* 9 (5): 27 (1989).

33. S. F. Schmidt, *Journal of Guidance and Control* 4: 4 (1981).

34. Shannon, *Bell System Technical Journal* 27. R. E. Kalman, *Transactions of the ASME: Journal of Basic Engineering* 82D: 35 (1960). Gallager, *IEEE Transactions on Information Theory* 47.

35. Pierce, *IEEE Transactions on Information Theory* 19.

36. R. W. Bass, *Proceedings of the IEEE* 84: 321 (1996).

37. www.nobelprizes.com.

38. E. O. Doebelin, *Engineering Experimentation* (Boston: McGraw-Hill, 1995).

39. J. D. Anderson, *A History of Aerodynamics and Its Impact on Flying Machines* (New York: Cambridge University Press, 1997). A. Pope and K. L. Goin, *High Speed Wind Tunnel Testing* (New York: Wiley, 1965). J. R. Hansen, *Engineer in Charge* (Washington D.C.: NASA, 1987). R. P. Hallion, *Supersonic Flight* (London: Brassey's, 1972).

40. J. D. Anderson, *Introduction to Flight,* 4th ed. (Boston: McGraw-Hill, 2000), p. 207.

41. J. H. Cowie, D. M. Nicol, and A. T. Ogielski, *Computing in Science and Engineering* 1 (1): 42 (1999). J. Heidemann, K. Mills, and S. Kuman, *IEEE Network* 15 (5): 58 (2001).

42. R. W. Cahn, *The Coming of Material Science* (Amsterdam: Pergamon, 2001), p. 23.

43. W. G. Vincenti, *What Engineers Know and How They Know It* (Baltimore: Johns Hopkins University Press, 1990), chap. 4.

44. Cahn, *Material Science.* P. Ball, *Made to Measure* (Princeton: Princeton University Press, 1997). A. Cottrell, *MRS Bulletin* 25: 125 (2000).

45. C. S. Smith, *A Search for Structure* (Cambridge, Mass.: MIT Press, 1981), chap. 5.

46. National Research Council, *Materials Science and Engineering for the 1990s* (Washington, D.C.: National Academy Press, 1989), p. 28.

47. National Research Council, *Materials Science,* pp. 5–6, 28.

48. A. Briggs, ed., *The Science of New Materials* (Oxford: Blackwell, 1992). O. Port, *BusinessWeek,* February 25, 2002, pp. 130–131.

49. R. Phillips, *Crystals, Defects and Microstructures* (New York: Cambridge University Press, 2001).

50. J. E. Gordon, *The New Science of Strong Materials,* 2nd ed. (Princeton: Princeton University Press, 1976). To see how preexisting microcracks can improve crack resistance, try this experiment. Paper usually tears easily along a fold line. Poke a hole (a microcrack) on the line beforehand. You will find that it stops the tear, because the tearing force, previously concentrated on a point at the tip of the tear, is now spread around the longer length of the hole's circumference.

51. J. S. Langer, *Physics Today* 45: 24 (1992).
52. R. Bud, *The Uses of Life* (New York: Cambridge University Press, 1993). J. E. Smith, *Biotechnology*, 3rd ed. (New York: Cambridge University Press, 1996). E. S. Lander and R. A. Weinberg, *Science* 287: 1777 (2000).
53. M. Kennedy, *Trends in Biotechnology* 9: 218 (1991).
54. Another element is monoclonal antibody technology. In 1975 Cesar Milstein and Georges Kohler succeeded in fusing two cells to form a *hybridoma*, a hybrid cell capable of indefinite proliferation and secretion of large amounts of one specific kind of antibody. Besides therapeutics, it facilitates diagnosis by providing easy tests.
55. G. Ashton, *Nature Biotechnology* 19: 307 (2001).
56. D. Meldrum, *Genome Research* 10: 1081, 1288 (2000). J. Hodgson, *IEEE Spectrum* 37 (10): 36 (2000).
57. S. J. Spengler, *Science* 287: 1221 (2000).
58. J. E. Bailey, *Chemical Engineering Science* 50: 4091 (1995). C. F. Mascone, *Chemical Engineering Progress* 95 (10): 102 (1999). J. P. Fitch and B. Sokhansanj, *Proceedings of the IEEE* 88: 1949 (2000).
59. T. Reiss, *Trends in Biotechnology* 19 (12): 496 (2001).
60. Introduction to the special section "Systems Biology" in *Science* 295: 1661–1682 (2002).
61. G. Stephanopoulos, *AIChE Journal* 48: 920 (2002).
62. D. D. Ryu and D. H. Nam, *Biotechnology Progress* 16: 2 (2000). S. G. Burton et al., *Nature Biotechnology* 20: 37 (2002).
63. J. E. Bailey, *Chemical Engineering Science* 50; *Science* 252: 1668 (1991). Stephanopoulos, *AIChE Journal* 48. M. Cascante et al., *Nature Biotechnology* 20: 243 (2002).
64. R. Langer and J. P. Vacanti, *Science* 260: 920 (1993). R. Langer, *Chemical Engineering Science* 50: 4109 (1995). M. J. Lysaght and J. Reyes, *Tissue Engineering* 7: 485 (2001). L. G. Griffith and G. Naughton, *Science* 295: 1009 (2002). See also the special issue of *Scientific American* 280 (4): 59–98 (1999).
65. Bailey, *Chemical Engineering Science* 50.

7.   LEADERS WHO ARE ENGINEERS

1. C. B. Smith, *Civil Engineering Magazine* (1999), www.pubs.asce.org/ceonline/0699feat.html.
2. D. Hill, *A History of Engineering in Classical and Medieval Times* (La Salle, Ill.: Open Court, 1984).
3. V. Bush, *Pieces of the Action* (New York: William Morrow, 1970), p. 151; in *Listen to Leaders in Engineering*, ed. A. Love and J. S. Childers (Atlanta: Tupper and Love, 1965).
4. S. G. Thomas, *U.S. News and World Report*, April 10, 2000, p. 86.

5. Quoted by J. B. Rae, in *The Organization of Knowledge in Modern America, 1860–1920,* ed. A. Oleson and J. Voss, eds. (Baltimore: Johns Hopkins University Press, 1979), pp. 249–268.

6. The world's first university business school, the Wharton School at the University of Pennsylvania, opened in 1881. Only two more were added in the next two decades. The Harvard Business School and Northwestern's Kellogg did not appear until 1908, and even then Harvard termed its school a "delicate experiment."

7. *Fortune,* April 15, 2002, pp. 132—167. Rick Priory, CEO of Duke Energy, studied and started work in structural engineering. Kevin Sharer, CEO of Amgen, designed and developed nuclear submarines before sailing as chief engineer on the *U.S.S. Memphis.* Sam Waksal, head of Imclone, is known as the "socialite scientist."

8. L. Burton and L. Parker, *Degrees and Occupations in Engineering 1999;* NSF 99–318, www.nsf.gov.

9. Several surveys in the 1960s found that almost all college graduates among Germany's top managers held degrees in three areas: engineering, economics, and law. And the number of engineers roughly equaled the sum of the other two. Other studies have found that about 60 percent of board members of German manufacturing firms have engineering backgrounds. S. Hutton and P. Lawrence, *German Engineers* (New York: Oxford University Press, 1981). G. L. Lee and C. Smith, eds., *Engineers and Management* (London: Routledge, 1992).

10. L. F. Haber, *The Chemical Industry: 1900–1930* (New York: Oxford University Press, 1971), chap. 10.

11. T. A. Stewart et al., *Fortune,* 140 (10): 108, 195 (1999).

12. I. M. Ross, *Proceedings of the IEEE* 86: 7 (1998).

13. A. D. Chandler, Jr., *The Visible Hand* (Cambridge, Mass.: Harvard University Press, 1977). D. F. Noble, *America by Design* (New York: Oxford University Press, 1977), pp. 277–286.

14. C. von Clausewitz, *On War* (New York: Barnes and Noble, 1832), vol. 1, p. 86.

15. A. D. Chandler, Jr., *Business History Review* 39: 16 (1965).

16. A. P. Sloan, Jr., *My Years with General Motors* (Garden City, N.Y.: Doubleday, 1964), p. 248.

17. P. E. Hicks, *Industrial Engineering and Management,* 2nd ed. (New York: McGraw-Hill, 1994), chap. 1. M. A. Calvert, *The Mechanical Engineer in America, 1830–1910* (Baltimore: Johns Hopkins University Press, 1967), pp. 14–17. Chandler, *Visible Hand,* pp. 272–281.

Industrial engineering is much more than Taylorism. Frederick Taylor introduced time and motion studies to measure the performance of human manual tasks. Despite its publicity, his method met resistance from within the engineering community and without. ASME refused to publish his *Principles of Scientific Management* in 1911, rejecting its claim to science. Time and motion studies are still being conducted, but by technicians. "Time study practitioners are not industrial engineers, if that is the limit of their activity and expertise," wrote B. W. Saunders, in *Handbook of Industrial Engineering*, ed. G. Salvendy (New York: Wiley, 1982), p. 1.1.4.

18. D. O. Belanger, *Enabling American Innovation* (West Lafayette, Ind.: Purdue University Press, 1998), p. 13.

19. R. Locke, in *Managing in Different Cultures*, ed. P. Joynt and M. Warner, ed. (Oslo: Universitetsforlaget, 1985), pp. 166–216.

20. S. Y. Nof and W. E. Wilhelm, *Industrial Assembly* (London: Chapman and Hall, 1997), p. 11.

21. K. Alder, *Technology and Culture* 38: 273 (1997). M. R. Smith, in *Military Enterprise and Technology Change*, ed. M. R. Smith (Cambridge, Mass.: MIT Press, 1985), pp. 39–86. D. A. Hounshell, *From the American System to Mass Production: 1800–1932* (Baltimore: Johns Hopkins University Press, 1984).

22. The legend that Eli Whitney achieved interchangeability in the 1810s in rifle production has been refuted through painstaking historical research, starting from R. S. Woodbury, *Technology and Culture* 1: 235 (1959). Whitney sold the idea vigorously but failed to deliver. He got a contract to make 12,000 muskets in 1798, but delivery was spread out over eight years instead of the promised two and quality was poor.

23. Chandler, *Visible Hand*, pp. 75, 485.

24. Sloan, *My Years with GM*, p. 20.

25. J. B. Rae, *American Automobile Manufacturers* (Philadelphia: Chilton Co., 1959). A. D. Chandler, Jr., ed., *Giant Enterprise* (New York: Harcourt, Brace and World, 1964). Hounshell, *American System*, chap. 6. L. Biggs, in *Autowork*, ed. R. Asher and R. Edsforth (Albany: State University of New York Press, 1995) pp. 39–64.

26. Sloan, *My Years with GM*, pp. 4, 118.

27. *Automobile Facts and Figures* 13: 6, 12 (1950). For lean production see J. P. Womack, D. T. Jones, and D. Roos, *The Machine That Changed the World* (New York: Harper, 1990).

28. E. Toyoda, *Toyota* (Tokyo: Kodansha International, 1985), pp. 106–109. Toyoda is a family name meaning "rich-yield rice field." Toyota, the name adopted for the company, has no meaning in Japanese.

29. Sloan, *My Years with GM*, p. 44.

30. T. Ohno, *Toyota Production System* (Cambridge, Mass.: Productivity Press, 1988), pp. 18, 71.

31. Ford factory worker Jim Grayson, quoted in S. Terkel, *Working* (New York: Pantheon, 1972), p. 165.

32. R. K. Lester, *The Productive Edge* (New York: Norton, 1998), ch.2.

33. T. A. Kochan, R. D. Lansbury, and J. P. MacDuffie, eds., *After Lean Production* (Ithaca: Cornell University Press, 1997).

34. M. Ishida, in ibid., pp. 45–60.

35. Ohno, *Toyota Production System,* p. 71.

36. H. Brooks, in *Scientists and National Policy-making,* ed. R. Gilpin and C. Wright (New York: Columbia University Press, 1964), p. 76.

37. See http://www.ostp.gov/pcast.

38. W. Light and B. L. Collins, *Mechanical Engineering* 122 (2): 46 (2000). B. L. Collins, ibid. 122 (4): 86 (2000). *IEEE Communications Magazine* 39 (4): (2001) is a special issue on the standardization of information infrastructures.

39. C. E. Harris, M. S. Pritchard, and M. J. Rabins, *Engineering Ethics* (New York: Wadsworth, 1995), p. 221.

40. R. H. Vietor, *Contrived Competition* (Cambridge, Mass.: Harvard University Press, 1994). N. J. Vig and M. E. Kraft, eds., *Environmental Policy in the 1990s* (Washington, D.C.: CQ Press, 1997).

41. J. M. Peha, *IEEE Spectrum* 38 (3): 15 (2001).

42. E. Wenk, Jr., *Making Waves* (Urbana: University of Illinois Press, 1995).

43. R. Nader and J. Abbotts, *The Menace of Atomic Energy* (New York: Norton, 1977), p. 365.

44. Office of Technology Assessment (OTA), *Nuclear Power in an Age of Uncertainty,* OTA-E-216 (1984). www.wws.princeton.edu/~ota/. A. M. Weinberg, *Nuclear Reactions* (New York: American Institute of Physics, 1992). D. Hochfelder, *Proceedings of the IEEE* 87: 1405 (1999). J. G. Morone and E. J. Woodhouse, *The Demise of Nuclear Energy?* (New Haven: Yale University Press, 1989). R. Pool, *Beyond Engineering* (New York: Oxford University Press, 1997). R. L. Garwin and G. Charpak, *Megawatts and Megatons* (Chicago: University of Chicago Press, 2001).

45. E. L. Quinn, *U.S. Commercial Nuclear Power Industry* (2001), www.eia.doe.gov/cneaf/nuclear/page/nuc_reactors. See the special issue of *The Bridge,* Fall 2001, at www.nae.edu.

46. R. Moore, *EPRI Journal* 25 (3): 8 (2000).

47. OTA, *Nuclear Power,* pp. iii, 3.

48. R. E. Hagen, J. R. Moens, and A. D. Nikodem, *Impact of U.S. Nuclear Generation on Greenhouse Gas Emission* (2001), www.eia.doe.gov/cneaf/nuclear/page/analysis/ghg.pdf.

49. National Energy Policy Development Group, *National Energy Policy* (2001), pp. 5–17, www.whitehouse.gov/energy.

50. D. Talbot, *Technology Review* 105 (1): 54 (2002).

51. J. Flynn, P. Slovic, and H. Kunreuther, eds., *Risk, Media, and Stigma* (London: Earthscan, 2001), chaps. 6–9, 13, 17, 21.

52. American Cancer Society, *Cancer Facts and Figures 2002*, pp. 35–36, www.cancer.org.

53. Except the faulty Chernobyl-type design, nuclear reactors usually employ defense in depth, which consists of many layers of safety features; when one fails, the damages are contained by the other layers. The world outside the former Soviet bloc has accumulated 8,500 reactor-years and experienced only one accident with a partial core melt, at Three Mile Island. At Three Mile Island the safety defense in depth worked and no significant amount of radioactivity was released into the environment. E. O. Talbott et al., *Environmental Health Perspectives* 108: 545 (2000), and other scientific studies overwhelmingly find no detectable increase in cancer rates among the surrounding population.

54. See www.gallup.com/poll/releases/pr020314.asp.

55. L. Winner, *The Whale and the Reactor* (Chicago: University of Chicago Press, 1986), p. 5.

56. M. Dertouzos criticized "five myths of the information age" in *Scientific American* 277 (1): 28–29 (July 1977). See also his *What Will Be* (San Francisco: Harper, 1997). I have discussed philosophical hype in S. Y. Auyang, *Mind in Everyday Life and Cognitive Science* (Cambridge, Mass.: MIT Press, 2000), pp. 64–76, 482–483.

57. C. E. Shannon, *IEEE Transactions on Information Theory* 2: 3 (1956).

58. K. Shrader-Frechette and L. Westra, in *Technology and Value,* ed. K. Shrader-Frechette and L. Westra (New York: Rowman and Littlefield, 1997), pp. 3–11.

59. H. Salzman and S. R. Rosenthal, *Software by Design* (New York: Oxford University Press 1994), p. 11.

60. In *Technology and Value,* Shrader-Frechette and Westra based their indictment of technology on an OTA report: "The U.S. Office of Technology Assessment (OTA) claimed that up to 90 percent of all cancer is 'environmentally induced and theoretically preventable.'" Then they identified "environmentally-induced cancer" with "cancer-causing technologies." Incredulous peer reviewers would find something totally different in the cited report: OTA, *Assessment of Technologies for Determining Cancer Risks from the Environment* (1981), available at www.wws.princeton.edu/~ota/ns20/pubs_f.html. On page 3, the OTA expressly warns: "Studies over the last two decades yielded a variety of statements that 60 to 90 percent of cancer is associated with the environment and therefore is theoretically preventable. As it was used in those statements and is used in this report, 'environment' *encompasses* anything that interacts with humans, including substances eaten, drunk, and smoked, natural and medical radiation, workplace exposures, drugs, aspects of sexual behavior, and substances present in the

air, water, and soil. Unfortunately, the statements were sometimes repeated with 'environment' used to mean only air, water, and soil pollution."

Data from *Cancer Facts 2002,* published by the American Cancer Society (ACS), mostly agree with the OTA's assessment. Roughly a quarter of cancer incidents are caused by *internal* factors such as genetic, hormone, or immune conditions, which are often not preventable with current medical knowledge. The remaining three quarters are caused by *environmental and preventable* factors. Like the OTA, the ACS is careful to clarify the meaning of "environmental causes." Among environmental factors, smoking and alcohol drinking account for about a third of cancer deaths. Another third of cancer deaths are "related to nutrition, physical inactivity, obesity, and other lifestyle factors." Technology also exacts tolls. The OTA finds a less than 5 percent association of cancer with air and water pollution, comparable to a less than 3–7 percent association with exposure to natural radiation from the sun and cosmic rays.

61. On the average, it now takes only 16 units of energy to produce an industrial good that required 24 units of energy in 1970. N. S. Pierre, *BusinessWeek,* 194F-H (November 27, 2000). V. V. Badamis, *IEEE Spectrum* 35 (8): 36 (1998).

62. Wenk, *Making Waves,* p. 22.

63. See *National Vital Statistics Report* at www.cdc.gov/nchs/ for data after 1900; data from 1850 to 1900, for Massachusetts only, are from U.S. Census Bureau, *Historical Statistics of the United States* (1975).

64. S. Scheffler, ed., *Consequentialism and Its Critics* (New York: Oxford University Press, 1988).

65. J. P. Bruce, H. Lee, and E. F. Haites, eds., *Climate Change 1995* (New York: Cambridge University Press, 1995), chaps. 1–2.

66. *Challenger* contained badly designed O-ring seals that had worked previously in temperatures normal to south Florida. But it was doubtful that they would work in the unusual cold forecasted for the scheduled launch day in 1986. Uncertain, engineers recommended canceling the launch. However, NASA was under tremendous political pressure to go ahead and, complacent after a string of successes, it failed to fully invoke its crisis-decision structure. The remark of Senior Vice President Jerry Mason to Vice President of Engineering Robert Lund, overruling engineering's objections, is quoted in T. E. Bell and K. Esch, *IEEE Spectrum* 24 (2): 36 (1987), and D. Vaughan, *The Challenger Launch Decision* (Chicago: University of Chicago Press, 1996).

67. H. Hoover, *The Memoirs of Herbert Hoover* (New York: Macmillan, 1951), p. 132.

68. G. Stix, *IEEE Spectrum* 25: 76 (1988).

69. Augustine, "Simple Systems and Other Myths," talk given at MIT, September 7, 2001.

70. E. J. Chaisson, *The Hubble War* (Cambridge, Mass.: Harvard University Press, 1994), p. 229.

71. See www.onlinethics.org.

72. J. Morgenstern, *Journal of Professional Issues in Engineering Education and Practice* 123 (1): 23 (1997).

73. H. Petroski, *To Engineer Is Human* (New York: St. Martin's Press, 1982), pp. xii, 97.

74. E. Gloyma, *Journal of Environmental Engineering* 12: 812 (1986). W. W. Nazaroff and L. Alvarez-Cohen, *Environmental Engineering Science* (New York: Wiley, 2001).

75. Census Bureau, *Statistical Abstract of the United States 2001,* Tables 363, 366.

76. D. Press and D. A. Mazmaman, in *Environmental Policy,* ed.Vig and Kraft, pp. 255–277.

77. M. A. Hersh, *IEEE Transactions on Systems, Man, and Cybernetics: Part C* 28: 528 (1998).

78. D. T. Allen and D. R. Shonnard, *AIChE Journal* 47: 1906 (2001).

79. C. J. Pereira, *Chemical Engineering Science* 54: 1959 (1999). M. Goldman, *Chemical Engineering Progress* 96 (3): 27 (2000). J. H. Mattrey, J. M. Sherer, and J. D. Miller, ibid. 96 (5): 1 (2000).

80. V. Bush, *Endless Horizon* (Washington, D.C.: Public Affairs Press, 1946), p. 141.

APPENDIX A

1. Census Bureau, *Statistical Abstract of the United States 2001,* Table 593.

2. *Science* 288: 2127 (2000) compares pay hikes and regrets among workers in various fields.

3. National Science Board, *Statistical Abstract.* P. Meikins and C. Smith, *Engineering Labor* (London: Verso, 1996).

4. M. Hersh, *IEEE Transactions of Engineering Management* 47: 345 (2000).

APPENDIX B

1. M. Reuss, *Technology and Culture* 40: 292 (1999).

2. J. R. Hansen, *Engineer in Charge* (Washington, D.C.: NASA, 1987).